Why Peacocks?

AN UNLIKELY SEARCH FOR MEANING IN THE WORLD'S MOST MAGNIFICENT BIRD

Sean Flynn

SIMON & SCHUSTER

New York London Toronto Sydney New Delhi

Simon & Schuster
1230 Avenue of the Americas
New York, NY 10020

First Simon & Schuster hardcover edition May 2021

SIMON & SCHUSTER and colophon are registered trademarks of Simon & Schuster, Inc.

For information about special discounts for bulk purchases, please contact Simon & Schuster Special Sales at 1-866-506-1949 or business@simonandschuster.com.

The Simon & Schuster Speakers Bureau can bring authors to your live event. For more information or to book an event, contact the Simon & Schuster Speakers Bureau at 1-866-248-3049 or visit our website at www.simonspeakers.com.

Interior design by Joy O'Meara

Manufactured in the United States of America

1   3   5   7   9   10   8   6   4   2

Library of Congress Cataloging-in-Publication Data
Names: Flynn, Sean, 1964- author.
Title: Why peacocks? : an unlikely search for meaning in the world's most magnificent bird / Sean Flynn.
Description: New York : Simon & Schuster, 2021. | Includes bibliographical. | Summary: "An acclaimed journalist seeks to understand the mysterious allure of peacocks-and in the process discovers unexpected and valuable life lessons"— Provided by publisher.
Identifiers: LCCN 2020056708 (print) | LCCN 2020056709 (ebook) | ISBN 9781982101077 (hardcover) | ISBN 9781982101084 (paperback) | ISBN 9781982101091 (ebook)
Subjects: LCSH: Peafowl—Anecdotes.
Classification: LCC QL795.P43 F59 2021 (print) | LCC QL795.P43 (ebook) | DDC 598.6/258—dc23
LC record available at https://lccn.loc.gov/2020056708
LC ebook record available at https://lccn.loc.gov/2020056709

ISBN 978-1-9821-0107-7
ISBN 978-1-9821-0109-1 (ebook)

For my boys,
Calvin and Emmett

Contents

Part One

ROOSTING

Chapter One

The reason to have a peacock, I would have thought, is self-evident.

When you suddenly, and without any relevant experience or hint of prior interest, come into possession of one, it is understandable that people would be curious as to why. Yet they present the question in a way that suggests they genuinely cannot see what should be plainly obvious. I'm sure it was from exasperation that George Mallory finally said he was climbing Mount Everest simply because it was there.

So: because of feathers. That is the reason. And colors.

Because a peacock is a wondrously improbable apparition, ethereal, an avian experiment strayed from a misty place where pretty things are whispered about before being made fully real. Because looking at one makes you happy. Because Keats was right about truth and beauty.

Also because, in this particular instance, anyway: Elvis, too. And because the first gifts you give the woman you've already decided you want to marry are freighted with enormously high stakes, some, even, that you can't possibly recognize until many years have passed and then one afternoon there are peacocks in the yard.

We were both writers, she in New York and I in Boston, and we met in the usual way people did before smartphones and swiping apps, which was through friends. When we still lived in different cities, I would pick up Louise at the Back Bay train station and take her to my house in Swampscott, a hamlet on the edge of the ocean north of Boston. The drive between the station and the house was a dreary slog past the airport and the greyhound track and sketchy secondhand-car lots on roads clotted with traffic and squeezed by crooked buildings smeared with soot. "I'm wooing you," I told her on the first visit, which was true. "This is the scenic route."

I narrated the highlights as best as I understood them, which was mostly from writing about murder and thieving and the hoodlum idiocy of the decaying Boston mob. "I knew a guy who used to fix races," I said when we passed the dog track. "He said in a pinch you can just kick 'em in the nuts." I pointed into one of the denser neighborhoods beyond the guardrail. "My first story for the *Herald* was over there," I said. "Convenience-store stickup. Guy got pistol-whipped."

She was charmed by those anecdotes, I told myself, by those dark tales so casually revealed. There is an element of mythmaking in the early days of dating: You craft the best imagined version of yourself, she crafts hers, and if you're lucky, you believe each other long enough for those two stories to twine together and begin to unspool a new one. That, at the time, was my best version: seasoned crime writer.

Louise, for her part, liked to say that she was never more Southern than when she lived in her Brooklyn walk-up. She fried chicken for me in the eight-piece electric skillet her mother mailed from Tennessee, and she gathered fresh collards from the bodega on the corner to stew in salt pork. She slipped spears of pickled okra into glasses of gin poured from the bottle in the freezer. We would sit with our stiff martinis near the window to watch the curious foot traffic circulating through the brownstone across the street. "That's a brothel, isn't it?" she whispered one night—not in judgment, I was certain, but to hint that she, too, was aware of things in the shadows.

Halfway between the train station and my house, traffic often stalled us in front of a storefront the color of penicillin mold called the Green Spot that, from the outside, appeared to sell only three things: lobsters, spider plants, and plaster busts of Elvis. "Mob shop," I told her. "That's gotta be a front."

She smiled every time we passed it after that. "*Mahb*stas and *lahb*stas," she'd purr. "*Lahb*stas and *mahb*stas."

I bought her one of those Elvis busts for our first Valentine's Day. A gift under such circumstances has to be precisely calibrated, as it reveals both your intentions and how closely you've been paying attention. A plaster Elvis was playful enough, but it was also hollow, literally and figuratively, and with a coin slot cut through the top. It hardly suggested my depth of feeling. I needed ballast.

She had told me, late one night when the okra spears were almost gone, about the book that made her want to become a writer, Flannery O'Connor's *A Good Man Is Hard to Find*. I wasn't very familiar with O'Connor, but I knew that the title story involved an escaped convict and multiple homicides, which I took as one of those small illuminating details that suggested Louise and I were well suited for each other.

I tracked down a first edition and had it wrapped and ribboned and leaning against Elvis's cheek when she arrived at my house. I baked a chocolate cake, deflated and with frosting like spackle, and set that next to Elvis, too. I thought it important she knew at the beginning that I wasn't afraid to fail.

Almost twenty years later, Elvis is in her office, the chips in his hairline touched up with Sharpie. The book is on a shelf in our bedroom. When I notice it every now and again, I remember Louise in a soft blue sweater and the light of a fire, one knee drawn up to her chin and talking with her hands the way she does when she's especially enthused. She's telling me for the first time about the Georgia farm where O'Connor famously raised peacocks. "In Milledgeville, same town as the state penitentiary," she's saying, because by then all of our best stories led to crime. "It's where they used to keep the electric chair."

That was a nice detail, the chair. I'd forgotten about the peacocks, though. We never know which parts are going to be important.

Still, that's another way to answer the question.

A more immediate explanation is to say that we already had the chickens and that we got the chickens because the snake died. He was a ball python, a docile species native to sub-Saharan Africa and big-box pet stores, and the only thing Emmett wanted for Christmas that year and, frankly, for every gifting occasion since kindergarten. He'd start lobbying a few weeks before his birthday, resume after he'd finished his Halloween stash, pick up before Valentine's and again near Easter. The calendar provides a surprising number of days on which to give a boy a snake.

The odds were always against him. Louise is afraid of snakes, as well as other toothy, wormlike creatures, such as lampreys and eels. But Emmett was in third grade that year, probably old enough to keep a pet confined to a tank, and he had greatly improved his pitch. He stressed the word *ball* and curled his hands into a small lump so his mother wouldn't confuse it with one of those Burmese monsters that might eat the cat. "And ball pythons are really gentle, Mom," he said. "They're not dangerous like all the copperheads in the yard."

That was true. There are copperheads in the yard. We'd moved to North Carolina thirteen years earlier, when Louise was pregnant with Calvin, Emmett's older brother, and we now live in a slate-roofed farmhouse on a misshapen acre shaded by pecans and sugar maples. There's an old smokehouse a few steps from the kitchen door, an aging barn across a graveled drive, and, around back, an abandoned greenhouse that's been converted into my office. The neighborhood is not remotely rural—the crop fields that surrounded the house long ago were subdivided into tracts of single-family housing, and we could walk to the nearest Target if we put some effort into it—but from a certain angle, the property can appear to be a farm still. It helps the illusion that there is a paddock in front of the barn, in which the neighbors keep a pair of Nigora goats and a miniature horse they brought home in a minivan. His name is Chief.

The decision to trade a cottage on the north shore of Boston for a Potemkin farm in North Carolina was impulsive, which typically we are not. Louise and I are both magazine writers, a job that requires only a laptop and access to an airport, and we'd been musing about moving somewhere warmer and brighter before we had children. As it happened, her father was treated for cancer in Durham, and Louise flew down every few weeks to keep him company. In between chemo and cranial radiation, the two would amuse themselves by scouting old houses. No one was home when they stumbled upon this place, so Louise's father told her to stand lookout while he peeked in all the windows. She was rightfully hesitant, but the old country lawyer in him was convincing. "No one will mind me," he said. "I'm a bald old judge with cancer, harmless as they come." He poked around long enough for Louise to get nervous. Then he got back in the car and sat there for a moment, not saying anything, just looking at the barn and the sweep of the porch. "This is it," he declared. "The one."

In fairness, we'd only ever seen two copperheads. But Emmett made a persuasive point: If we were already surrounded by venomous serpents, what was the harm in caging a small, timid python in his room?

"Can you promise me a snake won't get out?" Louise asked about a week before Christmas.

"Nope," I said.

She frowned at me.

"I mean, I'd *hope* it wouldn't get out. The cat would kill the poor thing."

She frowned harder.

"We don't have to get him a snake," I said. "We never have before. We can steer him toward something else."

She let out a long sigh. We both understood perfectly well that a snake was inevitable. "Fine, let's get him the snake." Her shoulders twitched, a reflexive shudder. "Christmas always makes me do things I'll regret." That wasn't true, I knew, but the Flingshot Flying Monkeys of two Christmases past had been somewhat traumatic.

I bought a snake the next day from a chain pet store, along with a tank and all the recommended accessories. Cosmo—Emmett was certain of the name—was supposed to be from Santa Claus, so I set up his tank in my office. I put aspen shavings on the bottom, set a water dish in one corner and a ceramic cave in another, and, between those, balanced a piece of driftwood upon which Cosmo could bask. Plastic gauges stuck to the wall measured the temperature and humidity, and a timer switched between two heat lamps, white for day and red for night.

Unlike Louise and Emmett, I had always been agnostic about snakes. I found them to be neither scary nor particularly interesting, and Cosmo did nothing to nudge my opinion. He was a basic and unpretentious brown mottled with spots the color of toasted oats, and a thin strip of scales along his spine caught the light from his day lamp; if he'd moved, he might have twinkled. But he did not move. Except for his tongue swabbing the air, he remained motionless. In the wild, he could have been mistaken for a tidy pile of dirt.

When I went back to my office that evening, though, the light from his heat lamp threw a fingerling shadow toward the door. Cosmo was stretched out and mostly vertical, balancing on his back third and nosing along the top edge of the tank. He slowly lowered himself, moved to a corner, rose again. He was bigger than I'd thought when he'd been curled up, more than a foot long. He checked the entire perimeter, crawled over his cave, across the driftwood, down to his water dish, then back around to the top of the cave.

We stared at each other for a few minutes, or seemed to, anyway. For all I know, Cosmo was looking at his own reflection or a smudge on the glass. But a boy in the third grade could easily imagine Cosmo was focused intently upon him, *communicating*, even, like that boa constrictor Harry Potter busted out of the zoo. The thrill was immediate, bubbly, because that right there was the privilege of Christmas parenting: the ability, or maybe the hubris, to appropriate the bonkers joy your child will feel when he wakes up to what he wished for. The feeling is a sort of arrogant relief, both that you've accomplished something special and, at the same time, not miserably failed on the most important day in a kid's year.

On Christmas Eve, after we read *The Night Before Christmas* from a fragile spiral-bound pop-up book that my mother read to me and her father read to her, the boys always read aloud their notes to Santa Claus, inquiring as to his well-being and requesting no more than three gifts. Then they toss them into the fireplace, having been convinced that magic smoke is the fastest conveyance to the North Pole.

Calvin went first. He is two years older than Emmett and had long ago figured out the impossibility of an obese man delivering gifts to the entire world in a single night; our neighborhood alone, he'd calculated, would take the better part of an hour. But he was sentimental enough to *want* to believe and kind enough to indulge his parents and, for several Christmases prior, not whisper anything about it to his brother. He asked only for a fish tank and a scooter, both of which were already in my office, fully assembled and wrapped.

Emmett stood before the fire next, holding his note with both hands. "'Dear Santa,'" he began. "'For Christmas I would like a real, living ball python.'" He put an italic emphasis on the word *living,* as if Santa might get confused and bring him a stuffed toy or forget to feed an animal in the bustling months before the big night.

Louise shared a subtle, satisfied smile. We had achieved peak Santa. No left-field gift requests, no forgotten batteries, no tiny pieces to lose. We would be in bed by eleven.

I went back to my office after both boys were asleep and it was time to put presents under the tree. Cosmo's tank was glowing on the floor, but something was off, out of place. I had a twitch of panic.

It took a second to adjust: The thermometer was missing.

No, not missing. Misplaced. It was on the floor of the tank instead of the wall, faceup in the shavings directly under the lamp, its needle pointed at a proper basking temperature of 91 degrees.

I slid open the top, reached for the thermometer. There was a riffle in the bedding, something pale and ropy shaking off the wood chips. I finger-swept the shavings away from the thermometer's dial and found Cosmo's head beneath it, lolled back but eyes open and tongue flicking. The thermometer was stuck to his throat.

I tried to separate the two, the round thermometer and the snake, but immediately stopped: The adhesive didn't give a hint of releasing. My panic gestated quickly. Christmas is when parents atone for the accumulated disappointments of the year, make up for the excess travel, the cranky deadline sprints. Santa was supposed to deliver a snake in seven hours, dammit. This was not the time for a soul-crushing complication.

I sprinted up to the house and found Louise in the bedroom, sorting candy for their stockings. "There's a problem," I said as quietly as I could. Calvin's bedroom is directly above ours. "Shit. I think it's a big problem. Shit, shit, shit."

"What, what, what?" She hopped up. "Please tell me something awful didn't just happen."

I mouthed the word *snake*, grabbed her hand, and started pulling her out of the room.

Down at my office, she gawped at Cosmo on his back with a big round thermometer on his throat. It was a very confusing scene. "What did you *do*?"

"I didn't *do* anything. Stupid snake must've pulled it down on himself."

"He can't be dead."

"He's not dead," I said. "But we gotta get that thing off of him."

We had a bottle of reptile spray—a concoction of aloe and emollients that I thought might loosen the adhesive—and a butter knife that could function as a dull scalpel and pry bar. But this was a two-person job.

"Oh God, do I have to touch him?" Louise scrunched up her face.

"You have to hold him, and I have to spray him."

She looked at the snake, looked at me. We had deeply serious expressions. Christmas depended on our skill with a kitchen utensil and a spray bottle.

"Okay," she said. "I can do that. Wait, do I have to do that? Seriously? No, right, it's fine. I can hold a snake. Of course, yes, I can do that."

She took a calming breath and reached in. Cosmo didn't have a lot of fight, so Louise just had to prop him up. I sprayed the joint between the snake and the adhesive until it was soaked, peeled the thermometer back a millimeter or two, doused everything again, waited for more of the sticky to melt. Louise had Cosmo almost completely upright at this point, and he seemed to be glaring at me over the big round dial.

"Does this hurt him?" she asked. "Oh, jeez. We're probably torturing him."

She was empathizing, I realized. Bonding, even, with a snake.

It took about twenty minutes, but the last of the adhesive finally let go. The thermometer took a few scales with it, but Cosmo wasn't bleeding, and when I picked him up for a closer look, he wrapped

himself in a lazy coil around my wrist. He wasn't obviously damaged, though when I set him back in the tank, he slithered directly
into his cave and hid.

We unplugged the lamps and the timer, and I carried the tank
up to the house, set it on a desk near the Christmas tree. Cosmo
was still hiding. "What if he's dying in there?" Louise said.

Oh, hell. *Dead snake* is pretty high up on a kid's list of worst
Christmas presents, like savings bonds and socks. We needed a
plan. If Cosmo was dead, two days would be enough to find a
replacement.

Louise does all the Santa writing, which is totally different from
her real handwriting because children are smart. Emmett's present
came with a nice note, written in Santa's loopy tweenage script:

> *Dear Emmett,*
>
>   *Because your snake traveled <u>very</u> far to get here, and be-
> cause this is a <u>new home</u>, he will need some quiet time to feel
> safe. <u>Do Not</u> handle him until December 27. This will give
> him time to learn your voice, the sights and sounds of his new
> home. I <u>know</u> you will take very good care of him—and he
> will be a good pet. Merry Christmas, Santa.*

On Christmas morning, Emmett was elated, as expected. He
was mildly disappointed in the embargo on handling Cosmo, but
in the short term the illusion was perfect. With the lights and the
driftwood and the shavings, it looked like there must be a snake in
there. If Emmett contorted himself to get his eyes even with the
floor of the tank and cupped his hands like blinkers on a horse, he
could see a dark lump that might have been Cosmo's snout. In the
meantime, not having a snake—*a real, living ball python*—to watch

gave him time to study the rest of the habitat. He focused briefly on the thermometer, which I'd prudently reattached at the very bottom of the tank.

"Why does that have the pet store logo on it?"

All the panicky adrenaline had kept Louise and me up; in the dark of our bedroom, we had worked through answers for the various worst-case scenarios, beginning, of course, with why the snake was dead (unpressurized high-altitude travel, obviously). Explaining away a logo was the least complicated. "C'mon, pup," I said. "Santa doesn't *make* snakes. They're living creatures—he has to get them from the same place everyone else gets them. He probably picks up the thermometer and stuff while he's there."

"Santa goes to the pet store?" he asked.

"Of course not," I said. "He sends an elf."

Emmett looked skeptical. He squatted next to the tank, his face level with Cosmo's cave, and blinkered his eyes again. "I think he moved," he whispered. "Yeah, he definitely moved." Louise and I left him like that, staring at a shadow, while we went to make Christmas breakfast.

Cosmo was not dead. He mercifully emerged early in the evening, throwing that same fingerling shadow up the wall, nosing along the edge of the tank the same way he'd done the day I'd bought him. The next morning, I carried everything up to Emmett's bedroom and put it on a shelf where he could see it from his bed. Emmett dutifully changed the water and checked the temperature and humidity. For weeks, Cosmo's entire behavioral repertoire consisted of hiding in his cave, lazing under a lamp, soaking in his water dish, or crawling from one to the other, a routine occasionally punctuated by unenthusiastically squirming around Emmett's wrist every three or four days.

The only other thing snakes are supposed to do is eat, which Cosmo refused to do. Knowing nothing of snake anatomy, I suspected that Louise and I had damaged his esophagus, or whatever equivalent it is that snakes have.

We kept a bag of tiny newborn mice in the freezer next to the Popsicles and the peas, and every five days Emmett thawed one under hot water and left it in the tank. Cosmo ignored it. Emmett moved Cosmo and the dead mouse to a dark box, and Cosmo ignored it there, too. Emmett sliced one open and wiggled it in Cosmo's face so he could pick up the scent of the entrails. Nothing. For two months.

And then he died.

I found him late on a Saturday morning when Louise and Calvin were out of town at a hockey tournament. Depositing some clean laundry in Emmett's room, I noticed that Cosmo was stretched out in the shavings instead of under his lamp, as was his habit. I gave him a little nudge. He did not move.

I called down from the top of the stairs. "Pup? I think Cosmo is, um . . . sick."

He pounded up the stairs. "What's wrong with him?"

"I'm not sure. You should probably come and look."

He looked for a very long minute. "He's just sleeping," he said. He looked closer. He looked from a different angle.

"Yeah, maybe," I said. "But I don't think so."

"Can we take him to the doctor?"

I was already Googling. There was a reptile vet thirty minutes away, and it closed in forty. "Yeah, if we hurry, we can get him there."

We dumped some LEGOs out of a plastic box, stabbed air holes in the top, tossed in a handful of shavings. Emmett reached for the snake. "I'll get Cosmo," I said. If he was dead, I could fake it, hold him so he wouldn't dangle, plainly limp and lifeless.

He was limp and lifeless.

I laid Cosmo in the box and started to put the lid on. "Wait." Emmett was holding the cave and the water dish. "He'll need these if he has to stay."

"Good point, pup."

Emmett rode in the back, holding Cosmo's box. For the entire drive, he was completely silent. I could see him in the mirror, staring at the box. He was pale.

We got out of the car in front of the vet's office. Emmett stopped before we got to the door. "Do you think he'll be okay?" He was looking at Cosmo, not me.

Cosmo was already dead. This was all a pantomime.

"I don't know, pup."

My cheeks flushed with shame. Emmett was for the first time facing the death of something he loved, and I froze. Death is one of the few things I should have been equipped to explain, too. I understand death. I'm *good* at death. Except in the clutch, when I lied to my son about his dead pet.

I write about death for a living. More specifically, I write about events, crimes and disasters and such, in which people have recently become dead. Magazine stories, mostly, long narratives about awful things that, over three decades and six continents, have involved many hundreds of dead people. I have no idea how many because I've never had the inclination to trudge back through the years and count them all. But in Emmett's lifetime, from his birth until the spring of third grade, the cumulative body count was four hundred and ninety-eight.

In fairness, almost half of those people were on a Malaysian airliner that disappeared in 2014. Most of the others were shot, which probably is the most common way people in my stories have died over the years, starting long ago with a newborn killed by his (clinically, legally, and temporarily) psychotic mother. A few were stabbed and some were blown up. At least twenty-five burned to death, possibly more depending on how finely one parses the cause of death, and the rest drowned or suffocated or were battered in one manner or another. One had a heart attack that almost certainly wasn't the result of poisoning, but opinions differ. One was sawed cleanly in two at the waist.

Sometimes there are forensic reports and legal briefs to help explain how those people became dead. But the actual mechanics usually aren't worth dwelling upon. Death is a plot point, and killing people on the page is not technically difficult, anyway. Death is an abbreviated narrative, the arc precise and clear, the trajectory wholly kinetic. Writing death well requires only restraint. There is no need to layer drama onto what is inherently dramatic, and no one deserves to have his or her last moment corrupted by clichés, especially nonsensical ones—bullets, for instance, are never *pumped* into anyone, though if they were, far fewer people would be killed by them.

The more interesting stories are who the dead were or what happened after they died or why they became dead in the first place or, usually, some combination of those things. Telling those stories, then, requires listening to the people who knew the dead or survived the catastrophe or treated the wounded or solved the crime or buried the bodies, and as many of those as are willing to talk. It's not an especially difficult job, that part. Anyone with empathy and patience could do the listening. Traumatized people want to talk to an interested stranger more often than you would probably suspect. They remember the most curious details, too, like how a skull looks when a bullet explodes out the back. They cry a lot, and you try to make yourself small in those moments, not to crowd their grief. As a rule, it is indecent to interrupt.

The harder part is absorbing all of those memories and all of that sadness and rage and distilling it into sentences coherent enough for a stranger to understand. That part takes more practice.

None of that, the listening and absorbing and processing, is remotely the same as grieving. But immersion and repetition are rigorous teachers. Strange, really, that I hadn't learned any wise and comforting words to ease the death of a snake.

The vet was about my age, graying hair and glasses, with a softness to his manner, the sort who will make a terrific grandfather someday. He introduced himself, asked Emmett his name, then nodded somberly.

"Emmett, I'm really sorry, but I'm afraid your snake died."

Emmett did not visibly react, except for the first glistening of tears that he managed to keep from falling out of his eyes.

"What was his name?"

"Cosmo."

"And how long have you had him?"

"Since Christmas." His voice cracked.

The vet gave a slow nod. "I see," he said. "And where did you get him?"

"Santa," I interrupted, possibly with a hint of desperation. There were only so many life lessons Emmett needed confirmed in one day.

"But I think he got him at a pet store," Emmett said. "All the stuff in his tank came from the pet store. And my dad said Santa can't make a snake."

"Well, yes, that's true. But I think Santa might have gotten you a sick snake. He hadn't eaten in a long time, is that right?"

"He never ate."

"Yeah, that means he was probably sick."

The room was quiet. "Maybe," Emmett said, "Santa shouldn't get animals from the pet store."

The vet vigorously agreed. I looked at my shoes.

We buried Cosmo on the east side of the silver maple, where the ground was soft and loose from all the seasons we'd tried, and failed, to grow asparagus and artichokes. Emmett made a tombstone from a broken slate that had fallen from the roof and planted it next to the grave. He kept the tank in his room so that he could look at it, and the view wasn't much different than the one he'd had on Christmas morning. If he wanted, and sometimes he did, he could pretend he still had a snake hiding in that little ceramic cave.

And then Louise got him the chickens.

She was buying tomato plants on the first Sunday in April at Barnes Supply Company, a shop west of downtown that an army colonel named Lee "Shorty" Barnes opened after fighting Nazis in World War II. It sold agricultural supplies until the spreading city absorbed the local farms, and then it sold mostly lawn and garden staples until Home Depot and Lowe's sponged away that business. So the George family, which bought it when Shorty retired in 1991, recalibrated it into pet supplies, food and toys and bedding and such, but retained the agricultural roots. Shelves of herbs and vegetable plants are wheeled onto the sidewalk out front each spring, and there are bins of seeds mounted to a wall inside. Bags of soil and amendments, compost and manure and whatnot, are stacked out back, and they stock the tools to sow, cultivate, and harvest a large garden. They sell feed for chickens and, in the spring, actual chicks. They're kept under lamps in the shed where the poultry waterers and feeders are shelved, with each breed, the Orpingtons and Easter eggers and Australorps, sorted into separate containers.

Louise brought home two barred Plymouth rocks, peeping charcoal puffs splotched with yellow. "The lady at Barnes said these are the best chickens to have as pets," she told Emmett. "I know they can't ever replace Cosmo, but they're pretty cool, right?"

Emmett, eyes big as walnuts, scooped up a chick. "I'm going to call this one Comet," he said, "and the other one Snowball." Calvin did not argue for naming rights, a solid big-brother move.

"Why Comet and Snowball?" I asked. "Neither one looks like a comet or a snowball."

"Comet does," he said. "A comet is a big dirty snowball in space"—ah, yes, he'd binged *Cosmos* twice on Netflix, which also explained the original Cosmo—"and the chicks look like dirty snowballs. But I can't call the other one *Dirty* Snowball."

"Fair enough," I said.

As luck would have it, chicks require almost the exact same environment as a ball python—a tank, shavings, and a heat lamp. They are, however, much more interactive than snakes. Every couple of days, we would cover the bathroom floor with newspaper, shut the door to keep the cat out, and let the chicks bounce around for a while. Chicken Time, we called it. They also grow very fast. Within a couple of weeks, Comet and Snowball were the size of squabs, their down replaced with feathers striped black and white, and ready to be moved to a small mail-order coop next to the barn.

I wasn't expecting much from them once they were outdoors. Though I've always been fond of animals, I held the same agnosticism toward chickens as I did snakes. So far as I knew, they were among the lesser avian species, flying only in short, panicked bursts and incapable of asking for crackers or repeating dirty words. They adapted easily to living brief, miserable lives on factory farms, which is not at all their fault but nonetheless a subjugation that is difficult to imagine, say, an eagle or a hawk tolerating. Once they stopped being cute, I assumed Comet and Snowball would be skittish accessories to the yard, like squirrels. They would lay eggs and eat ticks and fertilize the garden, all worthwhile contributions but nothing over which affectionate bonds typically are formed.

And yet they were such charming creatures. I was usually the first one up, so I'd release Comet and Snowball in the morning, which to a chicken is a time of great joy. Their heads would pop up in the window of their hutch when they heard me crunch across the gravel, and they would hop down with tremendous enthusiasm. Pure gratitude, chickens. They never stopped chattering, their jabbering clucks eventually softening into satisfied tuts. "How are the single ladies this fine morning?" I'd ask as I tossed feed on the ground. Oh, they had a lot to say. On weekends, they followed Louise and me to the porch, eating ants and spiders while we drank our coffee and read the news. One of them, and usually both, would flap up on the arm of a chair and, from there, into a lap or onto a shoulder. If they weren't with us and we couldn't see them, we just needed to call. "C'mon, chicken friends," one of us would yell, and they would come running in tandem, always together and perfectly synchronized.

By early July, the ladies were outgrowing their small coop. I decided to build something next to the barn, on the side where the roof extends like a giant carport over a wide patch of dirt where we keep the firewood and the lawn mower and, at the far end, a substantial pile of scraps and trash I kept promising to get hauled away.

I stared at the wall, puzzling out a new enclosure. Chief snorfled in the paddock. The chickens liked to wander near his hooves, and I used to worry he'd step on them, but he never did. It was midafternoon and very hot.

My phone chirped. It was a text to both me and Louise from our friend Tanja.

ANY CHANCE U GUYS WANT A PEACOCK? NO KIDDING!

I blinked once or twice.

I had never considered whether I wanted a peacock. When did that become an option? Where does one even get a peacock?

And why would Tanja think we'd want a peacock? Oh, right, because we have chickens and a barn, and it kind of looks like we live on a farm even though we don't and . . . Wait, Tanja's an attorney. How does she have a peacock to unload?

No, we didn't want a peacock. What would we do with a peacock? Where would we put a peacock? I'm trying to find space for chickens, and they're like, what, a tenth the size of a peacock?

My phone chirped again. Louise: I WILL SPEAK FOR THE GROUP: YES, PLEASE.

Chapter Two

The peacocks lived on a real farm the next county over, where Tanja stabled her daughter's horse, and about twenty miles from our faux farm. We turned off a blacktop two-lane onto a long driveway that curved around a pasture, through a cluster of outbuildings, and past a pen containing a fully ripe hog before ending in front of a pale two-story house surrounded by shade trees.

A peacock stood on the roof, his back to us. His train hung over the gutter like a ristra, catching speckles of midday sunlight sneaking through the trees. From our angle, the feathers were of no particular color, only sparks of green and gold, copper and turquoise, burgundy and blue-black, all of them flashing and fading and flashing again with the slightest movement, the whisper of a breeze, the bird shifting his weight.

It was the most magnificent creature I had ever seen.

Calvin and Emmett went running after more peacocks by the tree line, at least ten of them, maybe a dozen, one alabaster white, like a statue in the grass. The place smelled of pine sap and manure, and dust seemed to powder almost everything, the buildings, the chickens in the coop by the hog, the air.

Except the peacocks. They remained glittering, almost shiny.

Louise was staring at the bird on the roof. "What do you think?" I asked softly, almost reverentially. "Pretty spectacular, yeah?"

She didn't answer right away, just studied the peacock a little longer.

"It's bigger than I thought," she said.

Tanja had texted us only the day before, but the question of whether we wanted a peacock was presented as a matter of some urgency. Why wasn't made clear, but we were encouraged to get to the farm as quickly as possible, whereupon the proprietor, Danielle, would explain.

She appeared from out of the dust, a generation younger than I'd expected, with a sporty flush on her cheeks and blond hair pulled back loosely. She had an unfussy confidence that immediately struck me as required in a horsewoman, though I knew little about horses; Chief was the only one I'd spent a significant time around, and he seemed more like a very large dog than a horse. Danielle had dirt on her hands and mud on her boots and an understated swagger in her walk. She looked like she could wrangle a large animal and like she would use a word such as *wrangle*.

Peacocks, I would learn later, had been on Danielle's farm since 1977, shortly after her grandfather bought the place from the estate of a dead tobacco farmer. He paid cash for it, which Edwin Johnston was able to do because he'd made himself wealthy in the artificial-eye business. He was a pioneer, and perhaps the only practitioner ever, of traveling ocularistry. Ocularists make prosthetic eyes, which used to be glass but, since about World War II, have been custom-molded and hand-painted acrylic shells fitted over a base surgi-

cally implanted in the eye socket. It's a niche business that ten
to run in families, since the only way to become an ocularist is
apprentice under one, of whom there are extraordinarily few. Edw
married into the trade and learned it from his father-in-law, a thi
generation artisan who spoke not a lick of English but had root
Lauscha, a German glassblowing town and the traditional hom
the modern artificial eye.

Edwin opened an office in Parsippany, New Jersey, in the
1960s, but his business was mostly on the road. People lost
everywhere. Plus, existing prosthetics needed to be cleaned an
polished every now and again, maintenance that patients were i
likely to pay for if they didn't have to travel an inconvenient
tance. Edwin worked almost the entire East Coast, Connectio
Florida, crafting and polishing and painting fake eyeballs. Da
estimated that he earned, approximately, "a shit ton of money."

He came across the tobacco farm because it was near o
the cities on his route. Edwin and his wife, Eleonore, were
people, and he bought the farm intending to build stable
riding rings and to convert the fields into pastures and pad
When Eleonore moved from Parsippany with their four ch
she brought one horse, ten chickens, and a single pair of i
peacocks she happened to raise as a hobby.

At about the same time, a family from Rochester boug
land across the road. They had three children close in age to
and Eleonore's kids, the middle one a long-haired misfit
sort teenage girls often find attractive or at least intriguin
can imagine what happened," Danielle said. "It's the late
ties, there are all these kids getting drunk in hay barns, a
bam, here I am." There was a shotgun wedding, a short ma
quick divorce before Danielle was old enough to know her

had been married. Her father was an ironworker, and he'd go off to wherever there was a job, Texas or Oklahoma, and then come home for a month or two with a pocketful of cash. He'd run around with his daughter, take her to the water park, and then be gone again.

Danielle stayed on the farm with her mom and grandparents. Her main interest, and the farm's main purpose, was horses, but there were peacocks in the trees and on the roofs and scratching out dust baths in the dirt. The females nest on the ground, hidden in tall grasses and shrubbery, and one of Danielle's jobs as a kid was to follow them around to find out where they were laying. Occasionally, she'd find an abandoned egg that she'd crack on a rock baking in the sun to see if it would fry like on *The Flintstones,* which it never did, but mostly, she watched for chicks to hatch. When they did, she would catch them and put them in a pen until they were big enough to escape foxes and raccoons and the like. Other than that early intervention, the peacocks were feral: They weren't medicated or cooped or fed a special diet, though they would flock to the house when they heard Eleonore rattling a bag of Toasty O's on the porch.

Years passed, the peacocks a background blur for the horses. Danielle went off to college and the farm started to decline. As Danielle would tell me later, her grandfather was a religious man who believed the world would end before he left it of his own accord, and no one plans for a future he's certain will never arrive. Edwin and Eleonore lived in the present. They spent thousands showing Arabians, traveled the world ten times over, embarked once on a sixty-nine-day cruise. "He made that shit ton of money," Danielle said, "and blew every penny of it."

She came home for spring break in 2004, her senior year of college, and found a for-sale sign stuck in the turf at the edge of the

two-lane. Operating a horse farm requires money and labor, and Edwin and Eleonore didn't have enough of either, even with the land whittled down from the several hundred acres they'd originally bought to only sixty, part of which was sheltered from the taxman in a charitable trust. Danielle wasn't having it. She went to the end of the drive, pulled the sign out of the grass, and announced she'd stay and help run it.

She never did go back to college, but that's okay. She fixed the place up, got it looking so nice that neighbors would come by to see what the new owner had done and be surprised to find Edwin still there. Danielle adored her grandfather. He walked her down the aisle when she got married in 2007, and she held his hand in the hospital when he had cancer seven years later so he wouldn't die alone. She kept managing the farm until the estate was settled, and she eventually bought it outright with her husband, Doug.

There were about fifty horses on the property at any given time, most stabled, some Danielle's, and maybe two dozen peacocks. Predators killed a few of the birds every now and then, but chicks hatched every summer, so the numbers never appreciably dwindled. All told, there had been peacocks on Danielle's farm for three human generations and more avian ones than anyone had bothered counting.

"I have three over here," she told us, waving back behind a shed. We followed her to a little hut cobbled together from bricks and boards and scraps of goat wire. "There's two cocks and a hen in there."

The hen was on a high perch that lifted her into the wired-over window of the hut. She was crosshatched by thin shadows, but we had a clear view of her upper half. Her neck and head were predominantly white and splotched with emerald and traces of bronze, as if she'd been splattered with glitter paint. Thin white stalks sprouted from her scalp, each topped with a dot of fluff, like a tiny bouquet of Seussian flowers. "She's pied," Danielle said in a matter-of-fact way that assumed I knew what she meant, which I did not. "The males are regular India blues."

I moved two steps to the other end of the hut, bent down, peered through a lower panel of wire. Both males were on low roosts, trains sagging onto the floor, necks rising into shadow. A band of sunlight slashed across their breasts, which were an over-saturated sapphire, the color of fairy-tale lakes, and they appeared metallic, almost polished. The feathers on their wings, in marked contrast, were covered in muddy striations of brown and beige, the same as the pattern on Comet and Snowball, as if the wings were sturdy embankments to contain the blue, retaining walls to prevent it from leaking out.

"I want to keep these three together," Danielle said. "They're a social clique."

"That makes sense," I said with a bluffing certainty. "How much for the three?"

Negotiating in dollars for such specimens felt wrong, somehow embarrassing. We should be talking pouches of gemstones or magic beans.

"Maybe two hundred dollars? Let me think about it."

"So *three* peacocks?" Louise interrupted, directing the question more to me than to Danielle. She stepped back, swatted at the flies buzzing around her; she was standing next to a pile of horse manure the size of a yurt.

I turned toward Danielle. "Would we have to keep them penned up?"

"For a while, maybe six or eight weeks, until they figure out that's their new home. Once they eat out of your hand, they should be fine."

I straightened up, repositioned myself to look at the hen again. She was watching the empty sky, oblivious to or uninterested in the people staring at her.

Pied. I made a mental note to look that up. Also, *India blue.*

"So why are you getting rid of these, anyway?" I asked.

"They've been getting killed," Danielle said. "A great horned owl showed up, and it's been tearing their heads off. I've been finding decapitated peacocks all over the place."

"Really?" Louise and I glanced at each other, processing the image. "Just, like, everything but the head?"

"Yep. Heads torn right off. I guess owls like the brains or something."

"Owls get that big?" Louise asked. "Wouldn't it have to be, like, the size of Emmett?"

"Oh, they're huge! I saw it. I was coming home one night, and it swoops down and just stands in the middle of the road for a minute. It's like the size of a toddler."

Louise flicked her hand at another fly. I could tell she was trying to picture this boy-size owl. Probably wondering why we'd never heard about a menace of giant owls attacking labradoodles and children. That was beside the point: We were here to buy peacocks. "All right," I said, slapping my hands together. "We'll need a couple hours to talk it over, figure out if we've got room. Can we call you later?"

"Sure. But I want to find them new homes fast, so it's first come, first served."

I nodded. "Understood."

I did not understand. Peacocks were in frenzied demand? I'd only ever seen them from a distance in zoos, and now I'd stumbled upon this bustling, secret bazaar of exotic birds. It felt vaguely illicit.

Walking toward the car, I stopped, bent down. "Is it okay if we take a couple of these?" The ground was littered with peacock feathers as long as my arm.

"Please," Danielle said. "Take as many as you want."

Louise had spontaneously volunteered to take a peacock because a peacock, in a fundamental sense, is not a bird that one possesses so much as experiences; as with an especially moving work of art, the simple act of looking at it will stir emotions. A peacock, she imagined, would patrol the yard like a sentry in dress uniform, high-stepping through the irises and roosting on the low branches of the cedars or the high peak of the barn. Every so often he would throw up a fabulous spray of feathers for no other reason than to remind us that such a spectacle is possible. It would be inevitable and yet somehow a surprise every time.

That is what one peacock would do, but only one.

Louise did not want Flannery O'Connor's multitudes. She wanted a single peacock, a manageable number proportional to our small phony farm. The property was suitable for a pair of chickens, not a flock, after all, and the paddock was properly sized for a miniature horse, not a Thoroughbred. We were scaled for a solitary peacock, Louise insisted. Three was another matter altogether. A part-time job, she said. A petting zoo.

"You can't have *one* peacock," I told her on the drive home. "He'd be lonely."

"People have one dog, don't they? One cat. One snake."

Her logic was exasperating, but it did not change the fact that our peacocks had already formed a group, a bond. You can't just break up the band.

"And it'll be friends with the chickens," she continued. "Just like the chickens are friends with the goats and the horse—"

"Wait, wait," I said a little too gleefully. "We don't have one chicken. One chicken would be cruel. Because they're so social. Snowball and Comet need each other."

"True," she said. "But maybe peacocks aren't like chickens."

Twenty-four hours earlier, I hadn't wanted any peacocks, and for the same reason I'd never wanted koalas or a narwhal: The idea had never occurred to me. Now that it had, now that those fantastical birds had been presented as a reasonable proposition, *of course* I wanted one. A peacock, it seemed to me, was a flicker of happy imagination, an impossibly magical creature escaped from a dewdrop of unicorns and wood nymphs. The ones on Danielle's farm were almost transcendent, shining through the dust as if they were refracting light from a star we couldn't see. I felt a little sad for Danielle. Once she found new homes for all the birds, she would be left with a plain, boring horse farm.

"They won't roam," I said. "We'll put them in a pen. For now."

"We can't have three peacocks."

From the backseat the boys began cheering for three peacocks. "Let's get six peacocks!" Emmett shouted. "Let's get fifty peacocks!"

I leaned toward Louise and whispered, "They'll get their heads bit off if we don't take them." The boys didn't know this gruesome bit, and I didn't want to open up a conversation about decapitation just then. But it did seem important to remind Louise that this purchase wasn't an acquisition, it was a rescue.

"If we get fifty peacocks, you will have one less mom," Louise said over her shoulder. "But maybe we'll get a pair." She patted my hand. "So nobody's lonely."

Chapter Three

Before we could bring any peacocks home, two issues had to be dealt with immediately, those being food and housing. I assumed peacocks could get by on chicken feed for a few days. But where to put them? The inside of the barn was too cluttered, and the smokehouse too small. I would need to build something, which I was planning to do for the chickens anyway. But how big did it have to be? Would they need a lot of height, someplace high to roost? Danielle's hut was only a temporary holding cell. Surely three peacocks would need something more spacious.

I turned, as one does in such emergencies, to the Internet. I hesitated over the search term. Are peacocks confined in a pen? An aviary? I settled on the chickenlike "peacock coop," which is not, according to Google, an obscure search term. There were straightforward questions (*How do I build a peacock coop?*) and basic tutorials (*How to Build a Peacock Coop!*) and displays of both the twenty-two *and* thirty-four best peacock enclosures on Pinterest. I skimmed the blue hyperlinks until, halfway down the first page, glowing like a beacon, was an entry from Martha Stewart's blog called "Expanding the Peafowl Pen."

If Martha Stewart had a peafowl pen, it would be the best peafowl pen, which was obvious from her proper use of the word *peafowl*. To most people, there are only peacocks—boy peacocks and girl peacocks and baby peacocks. As a technical matter, however, there are peacocks, peahens, and peachicks, and they collectively are referred to as *peafowl*. It's an unpleasant word, what with that ugly *ow* sound, and there's almost never a reason to use it: No one has ever been awed by the beauty of a peafowl. Still, it is the correct word, and anyone dispensing pen-building advice should have a mastery of the nomenclature, even if I never expected to use it in conversation myself. I assumed, for that matter, that *pen* was the right term, but that it probably could be used interchangeably with *coop,* especially if one already intended to disregard the peacock/peafowl rule. Also, Martha Stewart in that post referred to her group of peafowl as a *muster,* which seemed to me much more dignified than calling them an *ostentation,* as some people do but which is cloyingly descriptive, like calling a group of frogs a *hopper* or a nest of vipers a *squiggly.*

That said, the language was all secondary. Martha Stewart's pen would be the best pen because she is Martha Stewart. I respect the Martha brand. She does not half-ass anything and she is a very good teacher, which I knew because I'd been down this apprenticeship road with her before: By God, Martha taught me how to make a *Bûche de Noël.* And I'm much better with a hammer than a jelly roll pan.

Martha's peacock residence resembled a miniature and impeccably kept pale gray barn. It had shiplap walls, a shingled roof, and thirty-foot runs of galvanized pipe and nylon netting extending from each side. A series of forty-two photographs chronicled two craftsmen, Pete and Chhiring, measuring and pounding and bolting and stapling one of the runs—the expansion referenced in the headline—until everything was square and level and sturdy.

I did not have a Pete or a Chhiring in my employ, and I had neither the time nor the money for galvanized pipes and shiplap. I did have long boards, though, a stack of half-rotted two-by-eights salvaged from old garden beds and piled under the shed roof of the barn, and folds of ancient chicken wire that were here when we bought the place. Close enough. If I cleaned out the rest of a pile of useful rubbish, I could build a pen against the side of the barn, under that part of the roof that jutted out like a carport, the same spot I was going to put a coop for Comet and Snowball. It wasn't level or square, but it was sturdy, with a metal roof and walls on the end and one side. Front to back was sixteen feet and the posts holding up the roof were ten feet apart, a footprint that would give three birds 160 square feet—not cavernous but better than the hut they were in. There was plenty more barn, too; I could always expand. And the ceiling was almost eleven feet—a handful of roosts at various heights would make it feel like a triplex.

Behind a broken lawn mower and eight broken shutters was an old screened door and, behind that, enough scrap lumber to build a frame around it. I screwed the long boards to the posts as a perimeter along the ground, and I used six more as stanchions rising to the rafters. The chicken wire was streaked with rust and creased in awkward diagonals, but I guessed there was enough to cover both exposed sides. My coop wouldn't be as attractive as Martha's pen—it would be a cage constructed literally from garbage—but it would be effective in the short term.

Five hours after we'd met, I texted Danielle. WORKING ON A COOP NOW. HOW MUCH FOR THE TRIO, AND IS A TOMORROW EVENING PICKUP GOOD?

Louise had compromised on only two, but I knew she'd come around.

Danielle replied immediately: WELL, I ALREADY SOLD THOSE 3 . . . BUT HAVE SINCE CAUGHT 2 MORE! A MATING PAIR MALE & FEMALE. $100 TOMORROW & YOU CAN HAVE BOTH?

Good for those three, I thought. Safe from the owl. And the price for two seemed like a deal. I'd never shopped for peacocks before, but fifty bucks apiece for mythological apparitions seemed like a steal. I texted back: SOLD!

The garbage coop was framed by the middle of the next morning, and the first run of chicken wire was stapled in place before noon. Comet and Snowball watched, softly clucking, from a sawhorse that was going to be repurposed as a temporary perch. Every so often, one of them would flutter down and scratch at an insect in the dirt. Farm equipment sheltered there for a hundred years had left an assortment of detritus—loose nuts from old machines, rusted nails, broken bits of clay pots, snipped ends of copper wire—that I'd mostly raked out the night before. In the process, I'd broken up the top half-inch of soil, which I assumed would be better for peacocks to walk on than a gumbo of sharp objects and loose gravel. Exposing a buffet of bugs was a bonus for the chickens.

Danielle texted me just before three o'clock, when I was putting the last screws into the spring hinges for the door. SO . . . PLANS CHANGED AGAIN. I HAVE 3 FOR YOU TO TAKE. WOULD THAT BE OK?? JUST LIKE THE FIRST 3 YOU LOOKED AT. AND . . . CAN YOU POSSIBLY COME SOONER?

She was unloading birds like Black Friday TVs. Clearly, I'd underestimated the demand for peacocks, which wasn't surprising, considering I hadn't known forty-eight hours earlier that there was a demand. In any case, we were back to three.

I replied that I'd be there as soon as I could.

FABULOUS!!! TEXT WHEN YOU ARE ON THE WAY!!

An hour later: BIRDS ARE BAGGED & READY TO GO!

That was a curious word, bagged. I imagined a peacock's head poking out of a grocery sack, his body restrained in some mysteriously efficient way yet his expression calm and regal, as if he were refusing to concede that he'd been bundled up like produce. They'd probably ride in the backseat, watching out the window like chauffeured royalty.

She wanted a bargain-basement $125 for the three of them. I told her I was leaving, just had to get some cash.

FANTASTIC! THEY ARE SITTING IN THE AC IN MY TRUCK. NO HURRY. She stuck a thumbs-up emoji at the end.

Danielle's pickup was running when Emmett and I pulled up next to it. The birds were definitely bagged, though less like groceries and more like gamblers who'd stiffed their bookie too many times. They were stuffed headfirst into feed sacks, the open ends of which were cinched with twine. The only thing poking out from two of the bags were legs zip-tied together and attached to unsettlingly large feet. This obviously was done for safety. A peacock's feet are the color of interstate pavement, scaly, and shaped like the claws in those arcade contraptions. There are three long toes in front and a short one in back, each tipped with a dagger of a talon. About an inch up the back of each leg is a spur that looks like a tooth from a large shark. I had not been aware that peacocks are so well equipped for violence.

The limbs poking out of the third bag, however, were barely noticeable because they were overwhelmed by the fully sprouted train of an adult male. The spine of each feather was bone white, and they appeared to be delicate stems in a dense iridescent bouquet. If anything, the feet seemed to be scraps of misplaced bark that Danielle forgot to pick out of a bundle of alien foliage.

"So this is how you transport a peacock?"

"If you've got the feed sacks, those are really the best," she said. "They're the right size, and the birds calm right down once they're in there and their feet are tied." She picked up the closest bag, handed it to me. I felt a bird shift inside. It was lighter than I'd expected. "This is the hen," Danielle said.

Emmett took two quick steps backward, as though the bird might explode out of the feed sack if he got too close. I laid the bag down gently, almost gingerly, in the back of our Subaru, trying to position the hen on her side. Emmett stepped forward again, leaned in for a closer look, snapped back. "Those feet are terrifying," he said.

"This one," Danielle said, holding the second bag, "is a juvenile male. And this big fella"—she jerked her head toward the feathers spilling across the seat of her truck—"is a black-shoulder male."

"What do you mean?"

She cocked an eye at me. "I mean he's got black shoulders. You'll see."

With the birds loaded, I paid Danielle and started down the driveway. I went slowly because I'd never had three peacocks bound and bagged in the back and didn't want to bounce them around. Surely they were fragile. They were definitely quiet. Ten minutes from home, I began to wonder if they were dead, or if they'd merely surrendered to their fate as cargo.

"Hey, Emmett, do me a favor," I said. "Reach back there and give one of those birds a poke. Gently."

I saw him in the mirror, furrows in his brow. "Poke it with what?"

"Your hand. Just reach over the seat and—"

"What if it kicks me?"

"It's not gonna kick you. I just want to make sure they're not dead."

"Dead! Why would they be dead?"

He looked stricken. Shit. I remembered too late that the last animal we'd had in a car was his dead snake.

"I'm sure they're not dead. Just nudge one. Talk to it."

He looked uncertain but started to twist around in his seat. I heard a soft crinkle when Emmett touched one of the feed bags, then an abrupt rustle. "It moved," he said. I was surprised that he did not immediately turn back. Instead, I heard another crinkle, longer. Emmett was stroking one of the bags. "It's okay, peacocks," he said. "You're almost home."

I pulled the car close to the garbage pen and Emmett bolted out of the backseat, hollering like a town crier. "We have peacocks! We have peacocks! Mom, Mom, come quick! We have peacocks!" He stood at the back of the car, not really hopping but sort of bouncing and flapping his hands as if shaking off hot liquid.

The birds were unloaded in reverse order. I carried the big male to the pen first. He squirmed, either frightened or annoyed that his surrender had been interrupted. I set him on the ground, still bagged and bound, and retrieved the next two. I realized I'd lined them up in a neat row, the way body bags are arranged after a fire or a nightclub shooting. I tugged the last bag out of the line, turned it perpendicular.

"That looks like three peacocks you've got there," Louise said.

"Shit," I said. She seemed to be suppressing a smile, but I couldn't quite tell because she was on the other side of the wire and backlit by the light of late afternoon. Last she'd heard, I was picking up two. Not three. "I totally forgot. It happened so fast. We had to go right away—"

"They'd better be spectacular."

The pen suddenly seemed quite small and the birds much bigger than when I'd put them in the car. And those feet. The talons appeared to have grown in the past thirty minutes, and the spurs were glinting like the edge of a razor. Danielle had told me to cut the zip ties around the feet, then start pulling the bags off. Odds were, they'd wriggle out before I could move the bags very far.

I closed the door and latched it with the hook I'd installed on the inside so I wouldn't have to worry about a bird busting out when I was puttering around in there. Emmett, Calvin, and Louise were on the other side of the wire, but I told them to take a step back. Who knew what havoc a freed peacock could wreak? Would the birds be angry and slash the closest non-peacock? Or scared and flap their wings and kick their feet, which as a practical matter would be the same as angry? And what happened when their feet were unbound? Had they been coiled like springs, waiting to erupt in a furious kicking spree?

It occurred to me that I really should have considered those questions before that moment.

I got down on one knee in front of the bag with the hen. I wrapped my left hand around her legs and carefully positioned the tip of a pair of tin snips over the zip tie. I clamped my left hand a little tighter, squeezed the snips, and heard the click of the tie severing.

Nothing happened. The bird didn't noticeably react. No kick, not even a wiggle. I loosened my left hand, and still no reaction. I let go completely. Nothing.

"Huh," I said, turning to Louise and the boys. "That was easy."

I scooted around to the other side of the bag, slid one hand under it to get the hen off the ground, and began to gently pull. The bird stirred, twisted to get her feet under her, and sort of backed out while I lifted the bag away. When she was completely out, she stood motionless in front of me. I realized in that instant, which was possibly far too late, that I had purchased three birds sight unseen, peacocks in a poke, as it were. But this one did not appear to have any obvious deformities. She was about the size of an affordably priced ham, and her back and wings were the color of stale chocolate. Her breast was a lightly toasted almond that darkened closer to her neck, which was short and woven through with green that increased as it got closer to her head. The green-brown continued up the back of her neck, over her scalp, and washed down to her beak. Her throat and the sides of her face were smudgy white, except for a slash of brown across each eye and dark dots on the sides of her head about where ears belonged. The overall effect was an almost calculated neutrality, as if a tangle of unremarkable underbrush had sprouted legs. She appeared perfectly camouflaged to nest on the ground.

She blinked at me once or twice, as if considering whether to move and how fast, whether I was a threat or an inconvenience. She settled on the latter and took a few casual steps to the far corner.

"What do you call that thing on her head?" Louise asked, waggling three fingers on top of her own head, a rough imitation of the tiny stem-and-puff feathers poking up from the hen's scalp. "A toodle?"

"Mom," Calvin said, an eye roll in his voice. "*Toodle* isn't a real word."

"She calls *me* toodle," Emmett piped up.

"All cute things are toodles," Louise said. "I use the word indiscriminately."

I pivoted to the next largest bag. "Okay, number two." I repeated the steps, but quickly, more confidently. When the sack was almost off, the young male ducked his head out and leaped toward the yard. He bounced off the chicken wire, careened to the back of the pen, ricocheted off the wall, and took another run at the chicken wire.

I stayed as still as possible until he burned himself out and settled in the back corner with the hen, of whose dignity I made a mental note. The male paced, two steps one way, two the other, over and over. He'd left cartoon dents in the chicken wire, like when the coyote gets blown through a wall, but he didn't seem hurt.

He had a toodle, too, except the puffs atop the stems were the same blue as his breast.

The big fella came out easy. A quick snip, an easy tug, and he stood up, stretched his wings, and calmly walked over to the other two. He had, as Danielle said, black shoulders or, more accurately, black wings. On closer inspection, it was more complicated than that. Instead of the brown-and-beige striation of the other male, his wings were a lacework of dark blues. The feathers were an indigo so deep it could almost pass for black except on the edges, where there was a fine and delicate thread of cobalt. He turned ever so slightly so that the light hit him at a marginally different angle, and the cobalt melted into the darkest jade. He shifted again, just a few degrees, and the cobalt returned. On his back, set off by the black wing on either side and the bright blue of his neck, were what looked like polished golden fish scales edged in a deep glossy green. His toodle was the same as the one on the other male.

The barn was very quiet, the four of us staring at this hijacked bird's extraordinarily intricate feathers.

"Okay," Louise murmured. "Pretty spectacular."

I nodded, still staring at the big one's back. "What do we call them?"

"The girl is Ethel," Louise said. "That's a good, sturdy name for a sensible bird." Calvin groaned and she poked him in the side. "If you'd been a girl, I wanted to name you Edith." He'd heard this before, of course, but it never failed to make him shudder.

"We'll call the young male Carl," I said.

Emmett gave us a sour look. "Those are dumb names," he said. "Why Carl?"

"No reason, really," I said, which was true. "I like how it sounds. One syllable. It's got that good *cuh-cuh* sound at the beginning. Goes with Ethel. Carl and Ethel, Ethel and Carl."

"We should call the big one Mr. Pickle," he declared.

Calvin burped out a laugh. "Why Mr. Pickles?"

"No, no—Mr. Pick-*el*. Just one pickle, 'cause there's only one of him."

"What does a peacock have to do with pickles?"

"His tail," Emmett said. "It looks like a giant pickle."

I looked at the bird. His long feathers lay on the ground, trailing behind him like an overlong cape. The shape was right, once you thought to look for it. The colors were all off, but the eyespots could pass for the bumps on a pickle. "All right," I said. "Mr. Pickle it is." It was no more ridiculous than a chicken named Comet.

Chapter Four

Early the next morning, after the chickens had been released for the day, I dug a mildewed camp chair from the recesses of the barn, unfolded it outside of the pen, and sat down with a cup of coffee to admire our peacocks. The boys and I had installed perches the evening before—a plank mounted kitty-corner on the walls; a pair of limbs, thick as rolling pins and cut from the privet that invades the periphery of the yard, that we hung with steel wire—so Carl and Ethel and Mr. Pickle, singular, wouldn't have to squeeze together on a sawhorse. I had no idea if they'd roosted on them: The birds had huddled as far away from us as possible while we worked, shifting together in a tight cluster and mirroring our movements from a safe distance. They remained together in a corner until it got too dark to see and we went inside for the night.

They were all on the ground again when I saw them in the morning, and they avoided me with the same efficiency. Before my chair was set up, they retreated as a group to the far side of the pen. But I could see there was less food in the feeder, and their water was down an inch or so. At the very least, they were nourished and hydrated, which I considered a small victory. Three peacocks had not died on my watch the first night.

"Good morning, pretty birds," I said. "How are you this morning?"

Ethel and Mr. Pickle looked at me. Carl stared at the wall. I wasn't expecting any particular acknowledgment, let alone the kind of enthusiastic greeting Comet and Snowball reliably provided. But I couldn't let them out to decorate the yard until they were comfortable with the barn being their home, which I assumed meant they had to be comfortable with me as well. Acclimating them to my voice and my presence seemed a logical first step. I could sort through emails and headlines while sitting by the peacock pen almost as comfortably as in my office. Plus, Mr. Pickle surely would hoist his enormous feathers at some point, and I preferred to be there when he did.

Comet and Snowball clucked at my feet, tilted their heads, looked up at me with greedy orange eyes. "Go eat some bugs," I said. "Appreciate your freedom. You could be like those big birds, stuck in a cage."

That word sounded unusually harsh to my ear. *Cage.* No, I reminded myself, this was a *coop*, a protective enclosure, sanctuary from a headhunting owl. On the other hand, I supposed, the difference between a coop and a cage depended on which side of the wire one stood. Those three birds had been running loose on a farm the day before, sun on their feathers, grass beneath their feet. That was all they'd ever known. Now they were under a metal roof instead of open sky, confined to a patch of dirt and pebbles the size of a FedEx truck, and saddled with ridiculous names.

"Straw," I told the birds. "I'm gonna get you guys some hay, something softer to walk on."

Nothing, just a blank stare from Ethel in the corner.

I drove to Barnes Supply later that morning, my first peacock-related excursion. I was a little giddy, stepping out as a rare breed myself, owner of magnificently extravagant birds. "I need a bale of wheat straw," I said to the first employee I saw, who was stacking sacks of high-end dog food.

"We got some out back," he said.

"Oh, good. I need it for my peacocks," I said with an overenunciated enthusiasm.

"Just one?"

"No, there're three, actually. I've got two males and—" I caught myself. His expression told me that wasn't the question. "Yeah, just the one."

"All right. Anything else?"

Another opening. "Well, yeah, come to think of it. What do you have to feed peacocks?"

"Game Bird Chow should do. Maybe the growth-and-plumage maintenance. Or you could just go with Flock Raiser pellets."

I pretended to contemplate this for a moment. Peacocks, I reasoned, probably were not exotic to people who sell game-bird chow by the pallet. Also, there's game-bird chow? And more than one kind? "Let's go with the growth-and-plumage," I said. "We'll see how my peacocks like it."

Back at the barn, I lugged the straw into the coop, scattered half on the ground, put the rest of the bale against the back wall, just inside the door, and sat down on it.

All three birds were at the front of the pen, reflexively keeping their distance from me. Only Carl was fidgety, pacing again, but he was up to three steps each way, expanding his range. Mr. Pickle was standing in profile to me and very still. His train popped with bright dots of sunlight but otherwise did not move. Only Ethel was looking at me. I thought she might be studying me, actually, that she seemed more calm and curious than edgy and wary, and then I thought that was a dumb thing to think because if I were her, I would definitely be more wary than curious. She was in a cage, and she'd gotten there after being jammed into a feed sack and zip-tied at the ankles. From her perspective, assuming she had one, I was her captor.

Danielle had told me the birds could be set free once they ate out of my hand, a conditioning that would take six to eight weeks. Training would commence immediately. Blueberries seemed about the right size, and I'd brought a carton into the pen. I pinched the top open, and Comet and Snowball started clucking. They were on a sawhorse on the other side of the wire and recognized the berries. Chickens have excellent daytime vision. "Stop," I whispered. They did not.

I picked one blueberry out of the carton and extended my arm like a sloth reaching for a leaf, moving slowly so the birds wouldn't startle. With the slightest flick of my wrist, I lobbed the berry toward Ethel. It landed in the straw about a foot in front of her. She didn't flinch, just looked at it, then returned to staring at me.

Comet and Snowball clucked more urgently. A blueberry was loose.

"It's a blueberry," I said to Ethel. She only blinked.

I tossed a dozen more before I slipped out the door. Ethel still hadn't moved, but the chickens rushed me. I sat on the sawhorse, fed them the rest of the blueberries. When I walked away, Ethel was picking at a piece of straw with the wheat seeds attached.

The next morning, I repeated the blueberry routine. The third morning, when the blueberries had been depleted, I switched to cherry tomatoes picked from the garden and cut into quarters. I was a bit smug about this because Martha Stewart had noted in a blog post that her birds "get lots of fresh, organic treats from my gardens." I had only one garden, but it was indeed organic. None of the birds appreciably noticed: Comet and Snowball ate organic tomatoes with the same greedy vigor as grocery-store blueberries, and Ethel and Mr. Pickle ignored both with the same frozen still-ness until I left. Carl pecked at the dirt.

After five days, I still hadn't seen Mr. Pickle hoist his feathers, a delay that was becoming unsettling. A peacock displaying his feathers couldn't possibly be a rare thing: I knew nothing about the physiology or behavior of the bird, but I had a decent grasp of its usefulness as a simile, and *that* peacock is forever flaunting his finery.

Granted, the clichés almost always are employed as a pejora-tive. To be as proud as a peacock is never intended as a compliment but, rather, always understood to imply vanity. To be as pretty as a peacock is not flattering, either, even if it's meant to be, because the mere invocation of the word hints that one is trying much too hard. (And to be told one "is a peacock in everything but beauty," as Oscar Wilde wrote, is just devastating.) As a verb, peacocking is straight mockery, though those being mocked typically are too busy peacocking to realize they should be offended. The Peacock Effect, meanwhile, is a psychological reference to the tendency of boys to show off in front of girls that arises in discussions of politics and marketing and schooling, and never favorably. We are taught even as children that the peacock is a difficult bird: Aesop repeatedly employed it as a reliable stand-in for arrogance and useless pride.

Implicit in all of those clichés is a peacock in full flower, and clichés work only if they're based on a common frame of reference. The reason "pretty as a royal flycatcher" or "proud as a puffer fish" never caught on is that hardly anyone can gather a mental image of what the industrious puffer fish should be proud of or what a royal flycatcher even looks like. But everyone can visualize, instantly, a peacock with his feathers splayed in an arc. It's a trademark silhouette, as recognizable as the Nike swoosh or McDonald's arches or, more obviously, the NBC peacock.

Yet no fair observer would accuse Mr. Pickle, his feathers dragging through the dirt, of insufferable pride.

I began to fret that he was a defective peacock. No, not defective: broken. That I broke him, if we're being honest about the situation. Broke his spirit, anyway.

Early on the afternoon of the sixth day—I varied the time, depending on what else I had to do—I settled onto the half-bale with a fistful of cut-up tomatoes and tossed a chunk toward Ethel. As soon as it landed, she stabbed her beak into the straw, plucked up the tomato, and swallowed. She wasn't skittish about it; she didn't appear to be keeping one eye on me in case I tried to grab her. It was as if she'd forgotten I was there, except then she looked at me. Expectantly, I thought, because I'd fallen into a habit of pretending that I knew what my birds were thinking.

She was about seven feet away. I tossed another piece, landed it a few inches in front of her. She picked it up, swallowed. Another one, not quite as far. She took a step forward, ate it. One more piece, a good eighteen inches short of her. She lifted a foot, looked from me to the tomato to me, and put her foot back down.

Mr. Pickle and Carl stayed far away, watching.

I considered this impressive progress for the day. I exited the pen, dropped the rest of the tomatoes for Comet and Snowball, and walked down to my office, a straight line across the grass and past Cosmo's grave. I wasn't there long, maybe a couple of hours, before I started back across the lawn.

A rustle caught my attention, the sound of a large desiccated bush being shaken. I stopped, tried to locate the source. The noise went off again, but it was more of a riffle than a rustle, more precise, cleaner, like a dealer shuffling a deck of cards with the edges gone soft.

Until that moment, I had no idea that peacocks made noise with their feathers. Even from thirty yards out and screened by chicken wire, Mr. Pickle was breathtaking.

I moved deliberately, torn between wanting to hurry and not wanting to spook him. The wire seemed to melt as I got closer, overwhelmed by the glow from inside the pen. Mr. Pickle turned toward me, his beak slightly open, as if he were mouth-breathing. His train was spread in a half-circle nine feet across, as high as my chin, and curling gently forward at the top. Except his feathers were no longer individual appendages. They were parts of a woven whole, an elaborate tapestry of gold and blue and turquoise. His breast and neck were a tapered sapphire wedge against the green-gold scales between his shoulders, which formed a smaller, denser half-circle, like a nova core exploding.

Mr. Pickle shuffled his feet, twisted a few degrees to the east, and the turquoise darkened to a deep jade. He turned back with tiny steps, and the turquoise returned. The top of his arc began to deflate ever so slightly, and then he rattled his feathers and the arc was full again. The entire train was alive, rippling like water, yet the eye at the end of each feather appeared to be floating on the surface, barely disturbed.

I was inches from Mr. Pickle, pressed up against the wire, but he did not back away. He was not startled. Quite the opposite—he was performing. He *wanted* me to watch. It was so unexpectedly intimate that I felt a prickle of embarrassment. But I couldn't possibly look away.

Chapter Five

We probably could have muddled along on our own, figured out peacock husbandry as needed. We'd done all right with Comet and Snowball. On the other hand, that strategy had proven suboptimal with Emmett's ball python. And Mr. Pickle's size and plumage suggested (admittedly without any evidentiary basis) that he was more complicated than a chicken, that peacocks had more moving parts and required more particular care. I decided to seek professional advice.

I searched "bird vet near me," and the first result was the appropriately blunt thebirdvet.com. His name was Dr. Greg Burkett, and his website identified him as a "board-certified avian veterinarian." His picture was at the top left: He was a cheerful-looking bald man in his fifties wearing a white lab coat over blue scrubs. A little bird with green wings, a white breast, a yellow throat, and a black head perched on his finger, trained well enough to turn toward the camera. On the right side of the home page was a head shot of a white duck with a bill that appeared to be shaped from putty. "Donate today," the caption read, "and help Dr. Burkett apply prosthetics for birds in need!"

That was my man right there. I called and made an appointment for the following week.

I hadn't recognized the address, but it turned out that I'd driven past Dr. Burkett's office more times than I could count over the years. I'd even used it as a landmark. "Turn right at that weird bird place," I'd say when I gave people directions, because that's what it appeared to be, a weird bird place. The building is a squat craftsman bungalow set back on a corner lot in the deep shade of oak trees, a kind of permanent dusk that can trick the eye into seeing warped clapboards and sagging beams, a tint of mildew. For years, there was a sign at the edge of the road, a red parrot bulging from a large square of molded white plastic gone yellow with age, that identified the place as "The Birdie Boutique: A Parrot Lover's Paradise." But there was no mention of veterinary services, let alone of cutting-edge avian medicine. I had always imagined a woman of a certain age doddering about inside, rearranging bird toys and birdseed and bird cages in the parlor, waiting for an occasional customer to peer in from the porch but not really caring if one ever came.

Up close, the building appeared to be in fine shape, a cottage reconfigured into a medical office and bird-supply shop. The former living room was an ambiguously tropical reception area. A thatched awning, the kind you'd find at the entrance of a tiki bar, shaded what appeared to be a rack of bird treats, and a second one hung over the opening of a dark hallway. The wall behind the reception desk was a pale Carolina blue, and it was covered with a mural of a giant macaw. The bird was perched on a branch, and fronds and vines seemed to slip down the wall below the ceiling tiles. "The Birdie Boutique" was painted in a French-nouveau font next to the macaw.

Why Peacocks?

To the right, a weary, worried-looking woman fidgeted in a chair. On the floor next to her was a dirty cat carrier from which I could clearly hear the frustrated clucks of a chicken. To the left was the retail section—the toys and feed I'd always suspected were there— and a large cage with the same bird I'd seen perched on Burkett's finger on his website.

The young woman behind the desk noticed me looking at him. "He's a caique," she said with a calm smile. "His name's Elvis. See? Because of the black hair. And it looks like he's wearing orange pants."

I hadn't noticed orange pants in the photo. I leaned in for a better look. Elvis was clinging to the side of his cage, chattering, friendly. "Oh, careful," she said. "He's a biter."

I glanced over my shoulder to see if she was being serious and noticed the sign on the wall behind her: "Safety First! Julie has worked 3 days without being bitten!"

"I'm here to see Dr. Burkett about some peacocks," I said.

"And he is waiting for you," Julie said. "Just through that door."

When Greg Burkett was a kid, his favorite television program was *Baretta*, a cop show that ran for four seasons on ABC in the late 1970s. It starred Robert Blake as a plainclothes detective named Tony Baretta who, in what was not yet entirely a stereotype, lived in a cruddy hotel room and drove a rusted rattle trap of a car, in his case a 1966 Impala he called the Blue Ghost. Baretta seeded seventies pop culture with such phrases as "Don't do the crime if you can't do the time," and Blake won an Emmy for best actor in a drama after the first season. (Thirty years later, in perhaps the grimmest irony of cop-show actors, he was acquitted of hiring a pair of low-rent hitmen to shoot his second wife in the head, though a civil jury later found him responsible for her death and ordered him to pay her estate $30 million. An appeals court later halved it, but Blake still filed for bankruptcy.)

Baretta's sidekick was a bird. A Triton sulphur-crested cockatoo, specifically, which is a big white parrot native to New Guinea with a bright yellow tuft on its head. The bird's name was Fred, and Fred was why *Baretta* was Burkett's favorite show. "When Fred came on and he talked and he rolled over and he drank from the liquor bottle and he answered the phone, that was it for me," he told me. "I watched that show every week for that damned bird."

In fact, Burkett decided that he wanted to train other birds to do Fred-like things. He'd always wanted to work in the movies—his uncle managed a theater in Kinston, a little city on the North Carolina coastal plain where he grew up, and his cousin ran a drive-in in Columbia, South Carolina—and teaching birds how to do tricks, how to *act*, was about as cool a movie job as one could have short of being a star.

In the summer of 1977, between *Baretta* seasons three and four and when Burkett was fourteen years old, his father retrofitted the family van into a serviceable camper and drove to California with Burkett, his mom, and his sister for a vacation. One of the highlights was a tour of Universal Studios, where *Baretta* was filmed. Burkett saw the Blue Ghost, and he sat through a show at the Animal Actors School Stage, which was, as the name suggests, a trained-animal show.

There were at least two cockatoos involved. The trainers running the show told the tourists that those birds had appeared on *Baretta*, which was possible—a female named LaLa did the close-up Fred scenes, but the show used several other cockatoos as well—and, to a fourteen-year-old boy, completely believable. Because he wanted to teach birds to perform, Burkett made it a point to corner the bird trainer after the show to ask him questions. He remembered the bird man was tall, with thick gray hair, and he remembered what the bird man told him: "I hate this job. Same thing every day. Hate it."

"That was the gist of it," Burkett told me. "'I hate my job, I hate my life.'"

(That would not have been the Freds' regular trainer. Ray Berwick was a Hollywood legend who trained more than twenty-five thousand animals, by his count, including all of the birds in Alfred Hitchcock's *The Birds*. In the 1980s, he wrote a book called *Ray Berwick's Complete Guide to Training Your Cat*, which, as anyone who's ever owned a cat will testify, could have been written only by a person with a deep and patient love of animals.)

In any case, Burkett was crushed. He spent the four-day return drive to Kinston in the back of the van trying to figure out a new line of future employment. Which he did: By the time the Chevy crossed the North Carolina state line, he had decided to become a bird veterinarian. It was a spectacular leap, from television cockatoo trainer to avian medicine. An epiphany, he said, a vision that was clear and sudden and unmistakably true. "I don't know why I had this idea I could doctor a bird," he said, "other than that they're living creatures and need someone to take care of them."

He mowed lawns and cleaned gutters until he had enough money to buy a pair of cockatiels from the pet store at the Vernon Park Mall. He named them Bonnie and Clyde and kept them in a cage in his room, where he expected them to breed so he could sell the hatchlings and start working his way up in the bird business. But he'd been sold a brother and sister; Bonnie laid only three eggs, one of which never hatched and the second of which contained a deformed bird that quickly died. The third, however, produced a healthy cockatiel that Burkett's father adopted as his own. He called her Pumpkin Cheeks on account of the daubs of orange on her face, and never caged her but instead let her perch on his shoulder as he went about his business. It was sweet, a grown man so infatuated with a bird.

With the cockatiels otherwise a bust, Burkett bought some budgies, which Americans insist on calling parakeets and which he'd seen performing in an old movie called *Bill and Coo*. They're reliable breeders, and he did all right selling baby budgies to people and occasionally pet stores, but he took a break when he went to college. By 1988, he was waiting to enroll in veterinary school and breeding birds in a basement. At the 1990 Great Smoky Mountains Bird Show, he met his wife, Missy (with whom he eventually would breed a lot of birds, fifteen thousand, ballpark, among thirty-five species); three years later, he graduated from vet school; and the year after that, he bought the house that he turned into an office. He had to evict the vagrant sleeping on a cot in the kitchen that's now the operating room.

Burkett was standing behind a stainless-steel table in what might have been a pantry at some point. He had an assistant next to him and notes on the table. He wasn't wearing the white coat, but he had on blue scrubs, and there was a silver lightning bolt in his left earlobe.

"So you've got peacocks," he said, more of a bemused statement than a question. He did not ask me why, either.

"Yep. Three. Two boys and a girl."

"Oh, that's not good. We'll get to that."

He asked me where I was keeping them and what I was feeding them. I described the coop, and he said it could be bigger but it wasn't dangerously cramped. I told him Barnes Supply had recommended Purina Game Bird Chow, and that I was surprised it existed, let alone that there were four varieties. My birds were on the growth-and-plumage maintenance formulation because . . . well, plumage, I guessed. That was a reasonable diet, Burkett said, though I should add fruit and greens and sunflower seeds.

"They seem to like blueberries," I said brightly, as if I'd thought of something else for his list. "And tomatoes."

"Um, yeah, they would," he said. "Those are, you know, fruits."

He had notes on parasites and worming and symptoms of illness, all of which he presented with a good amount of jolliness. "Now, if you notice a bird is sick, he's been sick for at least three days," he said. "They're very good at hiding it, so by the time you see it, he's really sick. So you should bring him in."

"Okay. So far no one seems sick. But when can I let them out?" I was hoping he'd have a more optimistic time line than Danielle, maybe tell me a month, tops.

He raised his eyebrows. "Never."

"What?"

"Never. Not if you want to keep them."

I frowned. This couldn't be right. "I was told a couple of months."

"Sure, you can let them out after a couple of months. And they'll fly away."

"Where?"

"Somewhere else," he said. "Looking for other birds. Especially with two males and a female. That should be the other way around. At least. Four hens for every cock is a good ratio."

Shit. Danielle had called my trio "a social clique."

"So I gotta keep these penned forever?"

"I would." An awkward silence followed. "Who told you a couple of months?"

I told him the whole story, leaning hard into the part about the savage owl.

"That didn't happen." He said it so flatly, so immediately, that I almost missed it.

"What? Of course it happened. The woman we got them from told us a giant owl showed up, and then she started finding birds with their heads torn off."

"I guess it's possible. But an owl usually wouldn't just bite off a peacock's head and leave the rest," he said. "A raccoon would, and probably just to be mean. Awful animals. Vile. But an owl would try to carry it away or stay and eat it. So would a coyote or a fox, if it could. If you only find some feathers, a coyote or a fox got it. If it disappears completely, that's an owl. If it's just dead, that's probably a dog that didn't know what to do with it once it killed it."

Burkett used to have four peacocks himself. Predators got both his hens, all of them, not just the heads. The black-shoulder male flew off, and he gave away the lone survivor, a white male. He knew the perils of owning peacocks firsthand.

"But she said it was an owl," I protested.

Another awkward pause. "Where'd you say you got these birds again?"

"A horse farm. Off 70."

"Woman named . . . Denise? No, not Denise. Danielle?"

"Yes! You know her?"

Burkett curled his mouth into a thin smile. He chortled, which I realized was an overused word when I heard a genuine chortle. "Hell," he said, "Danielle's been trying to get rid of those goddamned birds for years."

Chapter Six

In hindsight, it was apparent that Danielle was trying to get rid of her birds. Had we met under different circumstances, I would have recognized that, too. If, for instance, an editor had dispatched me to root around her peacock fire sale, I would have suspected as much within an hour, and I would have been all but certain by the next morning.

An owl? Seriously? For forty years there were no owls in the woods surrounding her exurban horse farm, and then one shows up that tears off peacock heads and leaves all the meat? That doesn't sound like a very sensible owl. I could have called an expert in avian behavior, probably Dr. Burkett himself, and he would have chortled at me that afternoon. He also would have suggested that two males and one female were not an unbreakable social clique but, rather, a sex-fueled cage match waiting to happen. And isn't first come, first served an odd way to sell any animal that isn't going to be eaten?

If I'd been working, I would have asked all of those questions. I'm curious by nature and tenaciously so when I'm getting paid. I would have been very pleasant about it, too. Aggressive interrogation is a lousy tactic if you actually want to learn something because it puts people on the defensive and no one should be defensive when discussing a topic as delightful as peacocks. Besides, I liked Danielle straightaway, and I still would have liked her when she stuck to her preposterous owl story. *More*, probably. She would have known that I knew she was bending the truth, and we both would have silently stipulated that it was a charming, harmless ruse, more of an interesting backstory she included for free with every bird purchase.

But I was not working, so I did not bother to consider that a woman who was finding homes for imperiled peacocks might not be entirely forthcoming.

Partly, that's because I have an unusually optimistic faith in human nature. Given the chance, most people will choose to be decent and kind, which, ironically, is something I've come to believe because I've waded through the aftermath of so much indecency and cruelty. It's a simple numbers game: I've spoken with many hundreds of people in the process of reporting many dozens of tragic stories, and only a tiny few were dishonest, malicious, or criminally inclined. The bad guys are outliers, the contrast to which everyone else can be compared: The Yosemite handyman who cuts off a woman's head throws into high relief the thirty-nine million Californians who've ever hurt anyone.

That's what I tell myself. Honestly, though, that whole theory could be horseshit. I might have come to the same conclusion about human nature while writing about municipal finance or minor league baseball. Or maybe I'm wrong. Maybe twenty percent of the population is three drinks shy of a murder spree, and I've mistaken sobriety and restraint for decency.

Whatever the reason, my faith in the basic goodness of most people is sincere, and my default inclination was to take Danielle at her word. But that does not mean I am a dope. I am not a passive vessel for self-evident codswallop (which, unlike *chortle*, is a word not used often enough). Burkett had not revealed anything to me so much as forced me into an admission: I didn't question Danielle because I didn't want to.

I did not want to know how long she'd been trying to get rid of her goddamned birds, and I did not want to know if she was white-lying about the reasons. What I wanted was a peacock. I wanted one of those glistening creatures to stake a claim to our yard, to be an elegant hallucination roosting on the barn, a cerulean sylph posing against the butter-yellow clapboards of the house. *Why* hardly seemed worth considering because, really, wasn't it obvious? A peacock is a thing of singular beauty. It was true that wanting one was impulsive and, possibly, greedy and selfish, but there was peril involved. Coveting a peacock is vaguely distasteful, but rescuing one, rescuing *three*, is noble, almost heroic. So, sure, tell me more about the head-ripping owl.

Probably shouldn't have gotten two males, though. I could see Burkett's point on that one.

Mr. Pickle and Carl did not fight. Burkett had told me that they probably wouldn't this time of year, seeing as how it was already the middle of July and the breeding season was over. Come springtime, though, they'd need to be separated unless I wanted to bring one in to get stitched up. Peacock battles are more performance than serious combat, he said, all that plumage flapping and flailing, but large birds are still jumping at each other, and sometimes a talon slices into a thigh or a breast. There'd also be more of a risk of that in a confined space, where they had less room to maneuver.

If Carl and Mr. Pickle ended up in one of these aggressive dance-offs, it would be comically one-sided. I doubted Carl would even have the confidence to put up a fight. Surely he must be as awed by Mr. Pickle's train as we were. Louise and the boys had seen it unfurled not long after I had but, oddly, never when they were together; the display was always for a solo audience, as Mr. Pickle seemed to worry that two or more humans might distract each other with their oohing and aahing, take the focus off him. Carl, by comparison, appeared to have stapled a worn-out duster to his backside. His train feathers were maybe a foot long and tilted at drunken angles, each one a few degrees off from the rest, and the colors were still muddy. It was an adolescent tail, the equivalent of pale fuzz on a teenage boy's lip, merely a promise of virility.

Carl was about two years old, maybe a yearling, definitely not yet three. I'd figured that out from some basic Googling I finally got around to—a peacock's train grows out in his third year. I was pleased to discover that I'd been using the right word: Collectively, those long feathers are called a train, not a tail. Individually, they are called coverts—because they cover other feathers, and near as I can figure, someone added an extra *t* to the word because *cover feather* sounded too simplistic.

The sprouting crown on the head, however, is not called a toodle. It is a crest.

Carl and Ethel were plain India blues, the autological name for the most common type of peacock, being that they are native to India and their breasts are that surreal blue (green peacocks, meanwhile, are endangered, and Congo peacocks are vulnerable, a slightly less perilous designation). Mr. Pickle was a marginally more exotic India blue: His black shoulders were a natural pattern mutation, sort of like red hair among humans. Pied, like that first hen I'd seen, is a bird splattered with white, which is a pattern mutation as well. Peafowl breeders had developed other mutations to produce birds with novel colors and patterns—midnight, peach, taupe, silver-pied, white-eyed, and so on—though I did not bother reading deeply into such obsession. How anyone thought an India blue, black shoulder or not, could be improved upon was beyond me.

Ethel was the least skittish of our peacocks, yet even she kept a practical distance from me, never coming close enough to be grabbed and stuffed in another feed sack. Time and blueberries, I assumed, eventually would close that gap. Someday they might even be as sociable as Comet and Snowball. Curious and fearless, the chickens would hop up on my chair and then my shoulder to pluck a ladybug from the brim of my hat. They tilted their heads when I talked to them and seemed to enjoy being handled in small doses. Emmett hugged them with such zeal sometimes that I worried he might break one of their tiny bones. What would Burkett think if the dipshit with the mismatched peacocks brought in a wounded chicken that had been hugged nearly to death?

Each night before dinner, one of us called the chickens to lead them back to their coop. "So they don't get killed," I'd remind the boys if I wasn't around to do it. "If they don't get closed in the coop, something will eat them." I usually pulled lockup duty, as I found the ladies to be pleasant evening company. They always came running when they heard my voice, a wobbly tandem of pure chicken giddiness. Most nights I had a treat for them, blueberries or kitchen scraps, and if I paused to pull a few weeds, they would frantically descend on the little dimple of fresh dirt I'd exposed and pluck out whatever unsuspecting insect had just been disturbed. If I detoured to rub Chief's nose, Comet and Snowball would stop at the fence with me. Then we would make our way to the peacock pen, next to which I'd moved the mail-order coop. The pen remained strangely quiet for a three-bird home. I would stand there watching while the chickens clucked around the perimeter, free and happy and chatty. Sometimes Louise or Calvin or Emmett would join me, and we'd silently stare at our strange new birds together, through the barrier of wire mesh, thinking the same thing, probably, though no one ever said it out loud: *Now what?*

After my appointment with Burkett, I told Louise what he'd said about not letting them roam free.

"So . . . they stay penned up forever?" she asked.

"I don't know. Maybe he's wrong. They were running wild on the farm."

We were in the kitchen cleaning up the dinner dishes. She topped off her glass of wine and leaned against the counter, thinking for a moment. "Did I ever tell you about my mom's pet duck?"

Yes, in fact, she had. I hoped she was bringing it up now in the spirit of finding a solution to a problem none of us wanted to name. And yet this particular story suggested a gruesome end for Carl and Ethel and Mr. Pickle; not every creative idea is a solution. As a child, Louise's mother had wanted a dog, but her sister had terrible allergies, so animals with fur were out of the question. Instead, one Easter she got a duckling and named it Cleo. By summer, Cleo was a full-fledged duck, and his owner, being seven, thought he might like to go on long walks in the neighborhood. She fashioned a special leash for Cleo, and they would take off down the hot pavement. It wasn't long before Cleo's tender webbed feet blistered. Here the story loses its specificity, but oozing infections were involved, and it ends with Cleo being dead.

"I think I've seen a picture of an emu on a leash," I said. "But we're not going to do that."

"Who knows what we're going to do," she said. "What does one *do* with peacocks, anyway?" She looked at me expectantly, as if she seriously thought I had something sensible to say.

"Well, I suppose—"

I had nothing to finish the thought. *Look at them in a garbage cage?* That was hardly a satisfying answer, for us or for the peacocks. I didn't *suppose* anything because, as was suddenly and uncomfortably apparent, I didn't really *know* anything. My preliminary research had begun and ended with asking Danielle how much she wanted for three birds. Hell, I hadn't even haggled. And I hadn't learned much more than feather basics since.

It was an excellent question, though: What does one do with peacocks? I could probably come up with an answer. My job, after all, is to tell true stories, to ask questions until I understand a particular subject or event well enough to explain it on the page. Surely I could, in time, explain peacocks to my own family. Researching the natural and cultural history would be no different than, say, deciphering the guidance system of a commercial airliner or the shifting alliances of warring tribes in Papua New Guinea. There would be travel involved—visits to people who do know what one does with peacocks—because reporting is always more fruitful when you show up. People are more open with a person who's come to visit than they are with a voice on the phone, and a Chilean desert or a Thai cave can be more accurately described if you've actually been there.

"I have no idea what one does with peacocks," I said. "But I think we can figure it out."

Part Two

MOLTING, MATING, AND MURDER

Chapter Seven

The morning was warmer than New York should be in October, and the sky was blue and clear, unseasonable weather that seemed to be a gift for the faithful and the curious on Amsterdam Avenue. There were many dozens of people waiting there, certainly hundreds, and they had with them dogs on leashes and birds in small cages and rabbits and hamsters and cats and turtles to be blessed in the Cathedral Church of St. John the Divine.

The cathedral is an Episcopal church on the west side of Manhattan north of 110th Street, and it is somewhat famous for blessing animals, which it does on the first Sunday in October because the feast of St. Francis of Assisi is October 4, and he is the patron saint of animals and the environment. The service was as much pageantry as worship. It featured dancing and giant airy puppets and music composed with the sounds of seals and wolves and humpback whales and, after the Eucharist, creatures and beasts processed through the nave: a camel and a horse and a miniature horse like Chief and ducks and geese and barred rock chickens like Comet and Snowball and a bright pink chicken and an ox and a donkey and, pulled on a cart, a tortoise the size of a pitcher's mound. A tortoise should be lethargic in autumn, but the temperature was pushing eighty degrees and had him all jumped up. An acolyte sat on his back to keep him from getting too frisky.

There was a single peacock in the procession, white as snow and cradled in the arms of a handler who followed a few paces behind two men with raptors on their forearms. A peacock was almost obligatory, as they are one of the other things for which St. John the Divine is somewhat famous. The one in the procession was a visitor, but St. John's has three of them—Harry, Jim, and Phil—who live on the close, which is what the church property surrounding the cathedral (as it is *close* to it, if you will) is called. Peacocks have been roaming the grounds since the nineteen eighties, when the Bronx Zoo and a Cathedral trustee donated four chicks, but the current three were gifts in 2002 from the graduating eighth-graders of the Cathedral School. The birds have the run of the place, and there is a shelter for them around back that resembles a miniature cathedral because it was designed by associates from the same firm that does the cathedral's other architectural work. St. John has architects, because almost 130 years after the cornerstone was laid, it still is being built, and probably will be centuries from now, too; monuments to God often are works in perpetual progress. As it is, St. John the Divine is already most famous for being one of the largest cathedrals in the world, a Gothic cavern of stone and flying buttresses. The facade on Amsterdam is more than two hundred feet wide, and the building runs more than six hundred feet to the east. The ceiling of the nave, 124 feet above the slate floor, is held up by granite arches the size of sequoias, and the air inside appears to generate its own atmosphere. There is a soft haze to the place, light sifting through stained glass from both sides and, from Amsterdam Avenue, ten thousand pieces of glass fitted into a rose window forty feet across.

Why Peacocks?

Before the service, Louise and the boys wandered the grounds with me looking for peacocks, which is fairly strange if you think about it too much—semi-wild peacocks cavorting about on the Upper West Side. But an urban church seemed as good a place as any to start sorting out the point of our own caged creatures, which was why I'd come once before and then brought Louise and Calvin and Emmett for the blessing of the animals. The boys had never been to New York, and a peacock hunt outside the world's largest Anglican cathedral seemed like one of the better reasons to miss a day of school.

Phil was easy to pick out in one of the gardens, as he is solid white, like the one in the procession. Not albino, which is a complete lack of pigmentation, but leucistic, a color mutation in the feathers that is fairly common among peacocks. We gave up looking for Harry and Jim; there were too many alcoves and nooks and secret peacock hideaways on an eleven-acre campus, and besides, we'd gotten distracted by the menagerie gathered along Amsterdam.

The Right Reverend Clifton Daniel III was slowly working his way along the sidewalk in a purple cassock. He is the dean of the cathedral, a tall, slim man in his early seventies with very little hair and a slightly crooked smile, and he appeared to be trying to personally welcome as many people as possible. He did not hurry. He has a deliberate, gentle manner, and when he said good morning or asked where you were from and why you were there, he did so at a pace and cadence calibrated to a calm, resting heart rate.

Someone asked him if there would be snakes in the service. "Undoubtedly," he said. All of God's creatures are welcome in the cathedral, and the serpent is no less one of His than is the fidgety tortoise in the nave or the white peacock in the garden.

* * *

The peacock has been entwined with the divine, with the eternal and the mystical, since as long as shamans and priests have been interpreting the universe through fantastical stories. How could it not be? The peacock's role in the epics of gods is a matter of form providing the function. To see a peacock, to look upon its blue breast and the thousand fleeting, shifting hues of its train, is to see, if one is so inclined, the pure work of the Creator. The peacock did not need to be invented from the imaginations of holy men and fabulists. It did not require mismatched parts, like the wings of Pegasus or the lion's body of the Griffon or the human arms of the golden Garuda, the mythological bird from whose feather, in Hindu lore, the peacock was created as a gift for the god Skanda. Even conjured from a feather, the peacock was delivered fully assembled, like the Milky Way or the sun.

In the older, Eastern traditions, the peacock was an accessory to the gods, and a malleable one at that. With his fierce claws and massive wingspan, he was a reliable mount for Skanda, the god of war. At the same time, Saraswati, the Hindu goddess of knowledge and the arts, was sometimes depicted riding on the back of a decidedly less ferocious peacock. In still other versions, Saraswati chose to ride a swan because the peacock was said to change its mind with the weather, thus representing indecision and fickleness, hardly useful qualities for either acquiring knowledge or conducting warfare.

In the *Uttara Kanda*, the last book of the ancient Indian epic *Ramayana*, Indra, the god of the heavens, took the form of a peacock to escape the rampaging ten-headed demon king Ravana. (In the folktale version, Indra just hid behind the peacock's fanned-out train.) "Pleased am I with thee," a grateful Indra told the peacock, which until then apparently was a large purple bird. "No fear shall spring to thee from serpents; and thy plumage shall be furnished with an hundred eyes." This is one of those fantastic stories built from real facts: A peacock is a nimble omnivore that often eats snakes. Having been given that ability by a god makes for a much better story, one that can easily be embellished and expanded. For instance, the peacock is believed in lore to be not only unafraid of snakes but also immune to venom, which it happily metabolizes into the iridescence of its plumage. Not surprisingly, peacock feathers have long been used in folk medicine as a curative for, among many other things, snakebites, a belief "so strong among the people of the Punjab that they smoke the feathers in a tobacco pipe as an antidote."

As travelers and traders spread the peacock west—King Solomon was importing them to Israel almost a thousand years before Christ, and Aristotle wrote of them as domesticated birds around 350 B.C.E.—the stories got weirder, darker. Simply explaining how the peacock's train came to be adorned with eyespots, for instance, could be a miserably grim tale.

In Greek mythology, the peacock was the favorite bird of Hera, who was a hot mess of a deity. She was the queen of the gods because she was married to Zeus, but she was also his sister who, as a baby—and this might be the root of the family dysfunction—had been swallowed by her father. Hera was beautiful but also jealous and spiteful, vain and cantankerous, outright cruel if she was in the mood, which she often was and which was not particularly surprising, because Zeus was a philandering cad. She caught him seducing a mortal named Io, but by the time she got down to Earth to confront him, Zeus had camouflaged Io by turning her into a cow. A very pretty white cow, but still, a cow. Hera, knowing the cow wasn't actually a cow, told Zeus she would be very pleased if he gave it to her as a token of his love. Which he did, because he couldn't think quickly of a good reason to say no.

Hera put the white cow out to pasture. She assumed Zeus would change Io back the first chance he got, so she sent Argus Panoptes, a giant with a hundred eyes, to keep watch. Zeus, being omnipotent, could have freed Io at his leisure, but then he would have faced the wrath of Hera, whom he feared more than he loved Io. After pondering this predicament for a while, Zeus decided to send his son Hermes to lull Argus to sleep and then, depending on who's telling the story, either sliced off his head or beat him to death with a rock.

With Argus dead, Io trotted off, and Hera sent a giant fly to bite her incessantly until Io escaped to Egypt and Zeus turned her back into a nymph and got her pregnant.

But back to the peacock, because this whole tawdry soap opera is meant to make a point about a pretty bird: Hera gave the peacock—*as a gift*—all of the eyes that were gouged from Argus. According to the ancient Greeks, then, the peacock is beautiful because he was decorated with the offal of a dead monster enforcer.

There is a thread of duality in the Western version of the peacock—the beauty always shaded with a bit of scorn, a negative tarnish for every positive. (The drab peahen doesn't factor into any of these stories, and the scriveners did not write about *peafowl*, either.) The bird is lovely but arrogant; or his plumage is gorgeous, but his feet are hideous; or his breast is an unreal blue, but his voice is that of a screech owl. There is in the peacock, always, a hint of the immoral, of damnation.

In one version of the Beginning, for instance, the peacock was extraordinary even by the standards of Eden. "He was the most beautiful bird of Paradise in voice and form, and he outdid them all in singing the glory of God," an Islamic storyteller named al-Kisa'i wrote almost a thousand years ago. "He used to go to the highest station of the seven heavens, whenever it came to his mind to do so, and his exaltation would reverberate throughout Paradise."

When the peacock wasn't exulting and such, his job was to guard the gate, which he did capably for a very long time. In this Islamic folktale, Adam and Eve wandered the fields for five hundred years, naked and unashamed, eating grapes and figs and pomegranates, and when they wanted to rest, they would retreat to their dais in the Dome of Contentment. They expected to remain in Paradise forever, so long as they obeyed one simple rule: Don't eat the fruit from the Tree of Eternity. This was not a vague instruction. God announced it quite clearly, and it's one of His more famous proclamations: It's in the foundational text for Christians, Muslims, and Jews.

Iblis, an angel who'd been cast out of heaven for not bowing to Adam, wanted to con the first humans into breaking God's one simple rule. The peacock found him skulking around the gate one day, asking to be allowed into Eden. Iblis told the peacock that he was an especially devout angel who'd been so busy praising God that he'd never glimpsed Paradise. If the peacock would let him in, Iblis promised to teach him three secret words. "Whoever says these words," Iblis said, "will never grow old or ill and will never die."

Now, it should be noted that one of the standard benefits of Paradise is eternal life, presumably without getting sick or old because that would make living forever more of a curse. The peacock really should have known this. Also, he should have wondered why an angel who claimed to be perpetually praising God was outside the gate chatting up a peacock.

The peacock agreed to let him in. He enlisted the serpent, who at the time was more of a psychedelic camel than a snake, to carry Iblis into Paradise, whereupon he tempted Eve to eat the fruit from the Tree of Eternity. Humans were cast out to suffer and die and be ashamed of their nakedness. The serpent was dragged about by angry angels until her limbs wore away and her body was stretched, and then she was sent to slither away on her belly. Then the angels turned on the peacock. "Leave Paradise forever," Gabriel told him. "So long as you shall live, you shall always be accursed."

Which was not unreasonable, given the circumstances. Dooming humanity to suffering and death and sin and shame is worthy of being accursed. Still, that's an awful lot to pin on a bird, even in a folktale. And odd, then, that the angels allowed the peacock to continue being beautiful.

Dean Daniel didn't really know much about the peacocks at St. John the Divine, but there's no reason he should. Peacocks are not liturgical creatures. They hold no significance in Christian Scripture, appearing in the Bible only three times and in passing, twice in accounts of Solomon's traders returning with them from far-off lands and the other when Job asked if God also created the peacock (He did). The bird is not a deity or even particularly holy.

(In one rare exception, the main deity of a syncretic faith practiced by the Yazidi is represented by a peacock. The theology is complicated, a mélange of mystic Islam, Christianity, Zoroastrianism, and other stray bits gathered in what is now Iraqi Kurdistan, and most of it isn't written down. Basically, the Yazidi worship the angel cast out of heaven, who they do not believe is Satan but, rather, a manifestation and messenger of God who has taken the form of a peacock; they are known among some scholars, in fact, as the religion of the Peacock Angel.)

Nor am I theologically conversant. I am a lapsed Episcopalian at best, a condition I have passed on to the boys. At the blessing of the animals, Emmett was alarmed by the incense, which he mistook for smoke and thus presumed the cathedral was on fire; and Calvin nudged me during a reading of the first verses of Genesis, when God is making the world in six days. "That's not true," he whispered.

"What's not?"

"What she just said. That didn't happen."

"Well, it's a metaphor—" I stopped. He'd had a closet creationist for a teacher once who'd told the class that Jesus made the ozone layer, which, in addition to getting both the science *and* the theology wrong, led to a discussion about the Bible not being literally true. The middle of a service in one of the world's largest cathedrals did not seem the appropriate time or place to revisit the matter.

Dean Daniel was kind enough to give me a tour of the close and help search for the birds. We spoke of the Christian iconography of the peacock, how the annual cycle of molting and regrowing an exquisite train was an obvious metaphor for the resurrection, how the early Christians adopted the coverts with the ocelli as a symbol of an all-seeing God. I'd read St. Augustine explaining how it was that souls could be tormented by hellfire for eternity without being consumed, the reasoning for which involved a slice of peacock meat he'd encountered in Carthage that, over the course of a year, did not rot. "For who but God the Creator of all things has given to the flesh of the peacock its antiseptic property," he wrote, which confirmed for me only that the peacock is a wildly versatile literary device.

Phil, the white peacock, was wandering near the statue of St. Francis. In his sermon during the blessing of the animals, Dean Daniel mused that St. Francis, were he to materialize on the Upper West Side, probably would nod politely at a statue of himself feeding animals but that he would be drawn to the statue nearer the street, the casting of a man sleeping on a bench called *Homeless Jesus*. He reminded the congregation that St. Francis renounced a life of wealth and privilege to minister to the poor and that maybe, possibly, the church bigwigs had anointed St. Francis the patron saint of animals to distract from that message. "I can't imagine that Francis would be comfortable in this world, or even perhaps in the church today," Dean Daniel said. "The call I issue to you is rooted in Francis's life and mission and ministry. I call on you, and me and all of us, to be kind. Simply to be kind."

He said that softly, in his calming cadence and with a trace of his native North Carolina basting his words, an accent that is subtly warm without the ear quite knowing why. I hoped the boys heard him clearly.

The dean and I found one of the other two peacocks—Harry and Jim looked pretty much the same to me—on a gate post. His train brushed the ground, long and luxurious, but he was not fussy about it. He appeared neither proud nor vain, only mildly curious.

Dean Daniel fed him some almonds from the stash he carries in his pocket. "They've created their home here just by being beautiful," he said. "And that's all they need to do."

Harry's neck, or Jim's, sparkled in the late-morning sun. Yes, a peacock really doesn't have to do anything more than that, sparkle.

"But it's amazing," I said, "that they're even here to do that. Big blue bird in the jungle, and those feathers hanging out like a handle for a jaguar or something. I mean, it's a miracle they survived long enough to evolve."

The dean didn't say anything at first, not that I was expecting an answer.

"Maybe a miracle," he said at last, "is really just life in God's kingdom slowed down enough for us to grasp it."

Chapter Eight

Mr. Pickle was spreading his feathers at least twice a day and more often three times. Considering I wasn't his primary audience, I assumed he was displaying more frequently, and I happened to catch sight of it on occasion. The average peacock spends about seven percent of his day with his train erect, which works out to just over four minutes of every waking hour. Mr. Pickle appeared to be skewing the curve.

Carl tried to display. He'd raise the scruffy duster on his behind and wiggle, shuffling his feet in a small aggressive circle. But the gap between the two peacocks was narrowing rapidly through attrition: Mr. Pickle had been shedding train feathers almost since Independence Day. I found one or two loose in the pen one day, then five or six by the end of the week. By the time the boys went back to school, the coverts were dropping in clumps and his actual tail, short and drab, poked out like a poorly designed prosthetic. At the end of August, it seemed that the only useful purpose of a peacock was to supply iridescent stems that could be gathered into mason jars and vases to give to friends' kids and our kids' friends.

Peacocks are only marginally more interactive than ball pythons—

ours did not like to be touched, let alone handled—and significantly less so than chickens, a behavioral disappointment the boys caught on to quickly. The novelty had worn off before the end of summer for everyone but me.

In fairness, the main reason Calvin and Emmet got bored with the peacocks was because of the puppy, a pug we named Tater because a pug puppy is the approximate size, shape, and color of a Yukon Gold. Unlike the chickens and the peacocks, Tater was thoroughly planned. Almost a year earlier, Louise had researched dog breeds, looking for one that was small but sturdy, accepting of grabby child hands, and possessing the temperament to endure being put on a skateboard or a trampoline or inside a dark pillow fort. In the spring, before the chickens, she hunted for a breeder until she found a hobbyist who kept meticulous records going back decades so she wouldn't inadvertently mate cousins and siblings and end up with puppies wriggling out of a cripplingly shallow gene pool. That Tater would be whelped in June by a pug named Louise was a pleasant coincidence.

For several years, our only pet had been a moody cat—a stray, as all our cats had been—and within a period of eight months, we had purchased a menagerie of seven animals, counting Cosmo. Yet the only one into which we'd put any real consideration was the dog.

Louise and I both had dogs when we were kids, and we always assumed we'd get one when we had the time and yard space that a dog needs. But we traveled too much, and cats kept finding us first. The current one in the house was a foundling that Calvin named Okra because that was where she turned up, the okra patch, ten years earlier. She was residually feral, tubby but quick on her feet, and she liked to bring us dead mice, sometimes half-dead mice, and once a baby rabbit. She was skittish and fickle and randomly peed on rugs and clawed couches, but we respected her mad predatory skills.

We assumed any dog we might have would come from a shelter or appear out of the mist one morning in need of a home. That's the kind of dog people we believed ourselves to be, which is to say, the kind of people we told ourselves we were: unfussy, trusting, open to all possibilities. Then we had children and, with them, a latent and wholly irrational fear that a seemingly docile shelter mutt might unleash repressed trauma in a furious, foam-flecked spasm of face biting if the boys picked the wrong moment for an overzealous hug.

So we got the pug. Ridiculous beasts, pugs, the playthings of ancient Chinese emperors and modern royals and the sort of people who might also enjoy, say, peacocks. Marie Antoinette had a pug. Queen Victoria had many pugs, as did Edward VIII and his wife, Wallis Simpson, the American socialite for whom Edward abdicated the throne. The pug was bred—no, designed—not to ferret out rodents or snarl at intruders but to amuse people, primarily by sitting on their laps. They are smash-faced and goggle-eyed and have such a shortened airway that they constantly wheeze and snort and grunt. Pugs also fart often and overheat easily, all of which suggest they are one of Dr. Moreau's moderately successful experiments. Was it ethical for us to encourage another litter of pugs, to throw cash money into extending that lineage? It seemed indulgent and, for the dog we wouldn't adopt, cruel.

On the other hand, none of that was Tater's fault. One can hardly blame a pug for not being a kennel mutt. Besides, it is impossible to be conflicted about a pug in the presence of one. A pug is the sort of dog for whom, when you take him to Barnes Supply for a harness, the staff will show you how to properly fit it so they can make an impossibly cute video of the whole thing for their Facebook page.

Tater became my companion on the morning rounds. He'd do his business, and then we'd release the chickens before visiting the peacocks. He tried to play with the chickens, but they scattered until they realized about a week in that there wasn't the remotest chance Tater might be capable of catching, killing, and eating them. I'd set up my camp chair outside the garbage coop, Tater would nap beneath the seat, and the chickens would scratch in the dirt near my feet while I answered emails and made calls. During breaks, I tried to teach Tater to fetch, but any stick big enough to throw was big enough for him to trip over.

One morning when Louise was working from home, she came out to find me but stopped before she got to the edge of the driveway. Maybe it was that I hadn't shaved in a while, or that I was wearing one of my summer hobo outfits, as she liked to call my comfortable clothing, or that I was in the mildewed chair that she had tried several times to throw away. Whatever the reason, she watched for only a moment or two, then went back inside. Later, she told me that between the denuded peacocks, the industrious chickens, the dozing pug, and my intermittent jabbering at all six of them, the dog was the least ridiculous creature in the yard. She said it as neither an insult nor a joke but as a reasonable observation, and one with which I could not in good conscience disagree. I *was* becoming a ridiculous figure, at least in the confines of my pretend farm with the organic garden and semi-exotic pets. It was almost intentional and would have been if I'd only admitted it out loud: I was creating, or trying to create, a home life as far removed as possible from the things I wrote about.

I'd begun trenching a gulf between work and home years earlier, when I realized the two were no longer properly siloed. Calvin was in second grade then, and he told me one night that a kid in his class was going to have him killed. Why he wanted Calvin killed wasn't entirely clear, but how it would happen was explained in detail: The boy said his father was going to come to school with a gun and shoot him.

Calvin told me this when he was already in bed, after we'd read *The Lorax* again but before I'd turned out the light. He was nonchalant about it, as if getting whacked by a parent were just an inconvenient possibility with which second-graders had to reckon. "He might tell his dad not to do it," Calvin said. "But he said if we hear any screaming from the bathroom, that's gonna be his dad giving Toby a beatdown."

"Toby?" That was a kid Calvin knew. "How'd he get dragged into this?"

"I dunno."

I waited to see if he had anything else to say, but he was quiet. "You know that won't happen, right?"

Calvin shrugged. Why would he know that? He'd been drilled for the day a homicidal loon would shoot up his school. He'd practiced for it, too, knew where in the classroom he was supposed to curl up until the shooting was over. It's a wonder he got out of bed in the morning, really.

I tried again. "Well, dads don't do that. I promise you. I know it sounds scary, but grown men don't go around shooting kids—"

"What about that man in Norway?" he said. The words came out fast, and the last syllables got tangled in the choke of a sob. Then he burst into tears, a great sudden deluge spilling out, as if a retaining wall in his tear ducts had collapsed.

I was startled by the ferocity. I pulled him close, a reflex, and felt him heaving against my chest. *Shit.* He wasn't supposed to know about the man in Norway. I'd always been vague with the boys about work, and for this exact reason. They knew I went places to write about things that had happened and that some of those things were sad and that sometimes people had even died. They would ask questions, and Louise and I never technically lied. But we left out a lot of details and deflected, we thought, masterfully.

Calvin began to calm down, his breathing slowing. He kept his head pressed against me. Norway had been months ago. I'd brought him a hockey jersey from Oslo, but I had no idea what I'd told him about the rest of that trip. He must have picked up snippets, overheard Louise and me talking through the story, plotting structure and pacing and tone. The man in Norway killed seventy-seven people: eight with a bomb in Oslo, sixty-seven he shot at a youth camp on an island in a lake west of the city, and two who died trying to get away, one by drowning and the other after falling off a cliff. I was in Norway about a year after the fact, during the man's trial, reporting what turned out to be an unusually long magazine story. I spent a lot of time with people who'd been on the island, who'd been wounded, who'd pulled panicked kids out of the lake. A police officer remembered being in a boat hours after it was over, hearing chirps and trills and snippets of pop songs from the cell phones that parents were calling and no one was answering. They were scattered all across the island, he said, blinking like fireflies. A man named Freddy told me, over several hours at a sidewalk café, how one of his daughters had called him from the island. She didn't say anything, just screamed for two minutes and seven seconds until the line went dead because the man shot her in the left side of the head and the bullet came out the right side and destroyed her phone. Freddy figured that out by studying photos from the place where his daughter was murdered.

Calvin was seven years old. I definitely didn't tell him that.

"What do you mean?" I said. "What about the man in Norway?"

"He killed all those kids."

Oh, hell, did I tell him the number? No, I wouldn't have done that. Probably not. Would it matter? Anything in double digits sounds like a statistic, doesn't it?

How long had *that* monster been under the bed?

I leaned back enough that he could see my face. "Yes, he did," I said. "He's a really, really bad guy."

Honesty seemed like a good approach. Acknowledge my son's fears. I was scrambling for the next line. What words make mass murder not scary? "There was only one of him," I said. "Do you know how many people there are in Norway?"

He shook his head.

Dammit. I didn't, either. "A lot. Millions." I was improvising. "After that one man—the one bad man—did that, thousands of those other people put flowers all over the city, giant mounds of flowers everywhere, because they were so sad and mad. There was that one bad guy, but everyone else tried to take care of each other."

"Okay?"

I guessed he was thinking the same thing I was thinking, which was: *So what?* The goodwill of the average Norwegian was useless in a North Carolina elementary school. Also, the dead people were still dead.

"What I mean is, there are very, very few bad guys in the world," I said. "And bad things happen, but very, very rarely. That's why I write about them, because they're so unusual."

He looked skeptical. His eyes were still moist.

"The man in Norway has nothing to do with your school," I said. "And this kid's father isn't one of the bad guys. Promise. Okay?"

Calvin nodded weakly.

He rolled on his side. His eyes were still open when I switched off the light. I stayed on the edge of the bed in the dark until he fell asleep, watching him and wondering what other terrible things I'd left in the shadows.

Louise and I were more diligent after that, always aware that one of the boys might be in earshot. We would preemptively explain what I was working on in slightly more detail, especially if it involved an event they might see on the news or hear about from their friends, but we dulled the sharp points, softened the focus. We would ask if they had any questions, but they rarely did. For the most part, my work to them was something that happened in faraway places. They did not need to know how much of it went on in my head.

In the weeks before we got the peacocks, I was writing about a dead soldier and his family. That task was not appreciably different from dozens of other projects, except I was having a terrible time of it, writing and deleting for days. Killing the soldier on the page wasn't difficult, though things I had learned about him when he was a boy reminded me of Calvin. Tiny details, quirks and mannerisms and such, but those are the ones that resonate; the soldier was familiar to me in ways that none of the other dead had been. And so I was dreading the paragraphs that had to come next. Unlike death, grief is extremely hard to write. It has no boundaries or shape, nothing physical to anchor the words. The temptation is to force it into view with adjectives and platitudes, and the danger is in slipping across the filament between moving and maudlin. Getting it right requires an intimacy that can't help but feel invasive, and it has to be right because it is so intimate.

Until the soldier, I had been able to stand outside of other people's grief. I could study it from a safe distance, close enough to see the details and feel the edges but still at a remove, as if from behind a sterile membrane. But because I could see my own child in the soldier, the barrier dissolved. Writing that father's grief, I finally figured out, was more visceral experience than intellectual exercise.

I discussed this at length with Comet and Snowball. I worked in the yard with them in the late spring, before the pecans had leafed out, moving my mildewed camp chair every so often to keep the sun off my screen and basal cells from sprouting on my pink Irish skin. The ladies would cock their heads while I read them sentences and paragraphs, staring at me with one orange eye each, not really listening, of course, but putting on a pretty good show of it. They looked like they were paying attention. When I got completely stuck, which was often, I sat in the grass and fed them blueberries. After a few days of this, I taught them to jump for treats. I was astonished that a chicken had a fourteen-inch vertical.

Louise found me on the lawn with Comet and Snowball one afternoon when she came home from her office. The chickens were leaping higher and higher, each patiently taking a turn. My laptop was open on my chair, and my notebooks were on the ground, a breeze rippling the pages. "Watch this," I said over my shoulder.

She knew I was late on a deadline. "Ah," she said with a pitying smile, "I see you've got yourself some therapy chickens."

I ignored her. She'd said the same thing about the bunny in the early spring. When I was stumped then, I'd sit on the steps and wait for the bunny to hop out of the boxwoods and into the irises, where I would toss him scraps of salad and watch him nibble until I wasn't stumped anymore. It was an efficient and mutually beneficial system until the night we heard a piercing shriek and saw one of the barn cats streaking away. My therapy bunny was under a boxwood with his belly slashed open. "It's a flesh wound," I insisted. Louise bundled him up in a towel and I drove to the all-night emergency vet, who thanked us for bringing in a young rabbit to be humanely put down, which wasn't my intent at all, but we were presented with no other options. We never mentioned the bunny after that.

The peacocks, within a couple of weeks of their arrival, had replaced Comet and Snowball as an audience for my more serious work conversations. They had the advantage of being physical captives, unable to wander away if they grew tired of me droning on, as the chickens occasionally did.

"If we start with the dead lawyer," I said to the peacocks one afternoon, "that's kind of backing into it, right?"

Ethel tipped her head to one side but did not betray an opinion.

"The airport scene with the live lawyer makes much more sense, I think. I mean, she's the one in the middle of all this."

Carl and Mr. Pickle stood a safe distance behind Ethel, shy slackers in the back of the lecture hall. They were excellent collaborators, never interrupting or criticizing, patiently listening to me sort out my thoughts, which was easier to do if I heard them out loud. Tater was much too excitable for such matters, and I felt less silly talking to a trio of large birds than to myself. Plus, the boys might always be close enough to hear me if I walked around muttering disjointed fragments about dead lawyers and airports.

All three peacocks watched me intently, as if expecting to hear something profound. They were actually expecting blueberries. Whenever I sat down inside the coop—I'd put a cinder block in there after the first bale of hay was scattered—they understood that I would produce blueberries or tomatoes or blackberries, whatever was abundant in the garden or cheap at the grocery. Yet they were still skittish enough to wait for treats, not cluck and hop and grab at my fingers like the chickens. It was easy enough to reframe their timidity as interest.

The tomatoes were always cut down to the size of large blueberries, and even some of the bigger blackberries were halved. I didn't want them to choke, which seemed to me conscientious and, more likely, moronic: Surely they were smart enough not to gobble things they couldn't get down their throats. Any species too stupid not to gag to death never would have evolved. Sometimes I'd leave a hunk of watermelon for them, a lavish gift that had Comet and Snowball stamping at the wire, and I'd scatter scraps of kale or the outer leaves pulled off of Brussels sprouts. They liked peanuts and sunflower seeds and dried mealworms. The only thing they refused to eat, for reasons I would never figure out, was strawberries. Even Tater ate those, green tops and all.

There was a limit to their patience, though. I rambled on for five minutes, maybe six, which is a very long time when you're rambling but not so long at all when you're trying to piece together a month of reporting. Carl got bored and looked out at the yard. Ethel took a single step toward me and stopped, a mildly insistent move, a soft demand for a blueberry.

"All right, then," I said. "We're all agreed, start in the airport, then circle back to the dead lawyer in the next section. Good?"

Ethel blinked.

"Yes, good." I pulled a blueberry out of the carton and balanced it at the tips of my fingers. Ethel watched it carefully, waiting. She knew I'd let it roll off into the straw soon enough.

None of them would eat out of my hand, though, which was slowing down my plan to release them into the yard. I was still hoping Burkett was wrong, that my peacocks were the exception to the fly-away rule. I'd read about peacocks staying on farms and in neighborhoods for generations. Martha Stewart allowed a couple of her big males to wander her property, and they hadn't escaped. Granted, she was working with 152 more acres than me. But my three supposedly were an established social clique—I was aware that I was being selective in the absurdities I chose to believe—so I thought there was a fair chance they might stick around.

We'd made progress. Ethel would march up as soon as I sat down, position herself no more than two feet away, close enough to touch, and wait. Mr. Pickle was running about two weeks behind Ethel, working up the courage to get close. The first time he moved in, I froze for an instant, then moved very delicately. Mr. Pickle and I were almost exactly eye to eye. He made no menacing movements, but from that angle, up close and sharply defined, clearly hungry, he seemed less like a regal bird of kings than an omnivorous jungle raptor. Neither bird would eat anything until I dropped it in the straw. Even Carl was getting brave, edging up behind Ethel and no longer startling if I threw a blueberry at his feet.

They were vocal by then, too. Nothing piercing, no screams or shrieks. There were soft trills and gentle clicks from Ethel that I took to mean she was content or curious, and a mild melodic hoot from Mr. Pickle, as if he were announcing himself at a cocktail party. Mostly, there was a deep honking, like gravel-throated geese, occasionally followed by a descending squawk. I heard it from Ethel initially one afternoon near the end of July. She had her neck puffed out like an oversized pipe cleaner and was staring into the bushes beyond Cosmo's grave. Mr. Pickle joined in, then Carl, all of them watching and honking and fluffing their necks. They were clearly sounding an alarm, but of what I had no idea; I could see only green leaves and dark shadows.

The honking would commence a couple of times each week,

never lasting longer than a few minutes. Sometimes they were directing it toward a squirrel foraging for spilled chicken feed and sometimes at the neighbor's cat sleeping under the forsythia. Those intruders were plain to my mammalian eyes. But peacocks, like chickens and most other birds, have different and in some ways much better daytime eyesight. They have extra cones in their eyes that allow them to perceive a broader spectrum of light, including ultraviolet, as well as a double-cone structure that detects brightness. Birds as a general rule also have superior hearing to humans. There's probably a continuous background of crunching and buzzing and crackling, and their visual world has colors that we can't see. But the advantage for a large bird that spends most of its life on the ground is that it can more easily see and hear predators.

The evolutionary trade-off, because there's always a trade-off, is that birds such as peacocks and chickens have terrible night vision. Peacocks in the wild roost high in trees, and the chickens in my backyard get locked up every night because they can't see what's coming to kill them. The reason is that they evolved with and from dinosaurs; the distant ancestors of chickens weren't hiding from *Tyrannosaurus rex* because they *were T. rex*. They weren't waiting for the big reptilian killers to fall asleep before scurrying around in the dark for food and sex. Mammals did that, which is why their eyes developed rod-shaped photoreceptors that function in low levels of light. Sixty-eight million years later, I can walk up on Comet in the night and she'll barely see me coming.

I never saw anything in those bushes, though. Ethel would start honking in that direction every now and again, and I'd wait and watch, thinking I might catch the jiggle of a branch, the slip of a shadow, some creature darting away. But nothing ever moved, nothing made sound, so there was really no point in worrying about it.

* * *

Mr. Pickle's train began regrowing almost as soon as we'd collected the last jar of feathers. By Thanksgiving, his coverts had grown out enough to almost cover his tail feathers. He displayed them about half as frequently as he had in July, and they fanned out thick and lush but very short, like a bonsai train. Carl's coverts had come in, too, but in an abstract and sickly sort of way. They were the blurry beige and black of the summer, albeit longer, with wide, airy gaps between them. He'd sprouted a single bright feather with an ocellus at the end, and it launched hard to the left, drooping like a broken antenna.

The coop was barely wide enough for Mr. Pickle's fully expanded train in the summer, and that's when it already was beginning to fall out. If Carl grew anything of note, there wouldn't be room for the two of them. Plus, Burkett had told me they'd need to be separated in the spring, and I couldn't split the pen as it was: The birds would each technically have enough space, but only in the way that prison inmates in solitary have enough space in their cells.

There were enough of the long boards left to double the size of the enclosure and, by incorporating the cedar posts holding up the roof, finally frame a separate coop for Comet and Snowball. It wasn't fancy, basically a chicken-wired closet, but big enough for their little mail-order hutch and the small table on which we kept a bucket of daffodil bulbs. The ladies had commandeered the bucket months ago as their laying spot, and we kept the little coop so they'd have a cozy space to huddle against the cold.

Which came on hard and fast. New Year's Day was seventeen degrees before sunrise, and the temperature dropped all week before bottoming out at seven degrees the following Sunday, with the wind making it feel like zero. I knew chickens and peacocks could tolerate the cold—they'd lived wild on Danielle's farm for decades—but teens and single digits were extreme in North Carolina. The day after the cold snap began, I started devising an arctic-weather-mitigation plan.

I needed a heat source, like what we had for Cosmo, only bigger. I couldn't have any open flames, obviously, and space heaters also were out. I could easily see Mr. Pickle's feathers catching fire, him running around in a panic, lighting straw and century-old barn wood on fire. Heat lamps probably wouldn't get hot enough to ignite anything, and I could suspend them away from anything flammable. The extension cords were all in good shape, and the entire run, from outlet to coop, was sheltered.

I went to Barnes Supply on the second day of January and bought two brooder lamps with clamps and two bulbs. I hung the lamps with steel wire and, directly below, set a board across a pair of sawhorses. I eyeballed the gap and was pretty sure Mr. Pickle wouldn't scorch his crest.

The front edge of the peacock pen was a wall of thin wire. A strong wind would overwhelm whatever heat my puny lamps generated. A windbreak of plastic sheeting would take at least ten degrees off the chill, maybe twenty. I got a staple gun and a utility knife, climbed a ladder, and began tacking up plastic. As darkness fell, my hands were stiff, almost numb, but I felt a pinch in the side of my forefinger, then saw a blot of red slide down the plastic. A trail of blood curled around to my palm. My finger didn't hurt, but the blood was smearing the plastic. I climbed down and went inside to wash out the wound and strangle it shut with three plastic bandages. When I came back out, the glow of the red lights behind the blood-smeared plastic gave the pen the look of a cheap and dangerous brothel.

Only half of it was sheltered, but it was a fine bit of jerry-rigging. From the inside, it had the cast of an overlit old-fashioned darkroom. I waved my hands under the lamps and felt a blush of warmth. I stood on the exposed side, then sidestepped to the wind-broken side, then back. "What do you think?" I said to the birds. They were busily picking through the fresh hay I'd bought when I got the lamps. They loved new hay.

The temperature kept dropping, and before I went to bed, I trudged out to the pen to see how the birds were adjusting. Comet and Snowball were snugged into their little hutch inside the big coop, their only complaint seeming to be that I'd disturbed them.

But in the peacock pen, the perch beneath the lamps was empty. All three birds had moved into the exposed half, the colder half. Mr. Pickle was on a plank in the back corner. Carl was on a branch in the middle. And Ethel was perched up front near the wire, wind ruffling her feathers, as close to a polar vortex as she could get.

Chapter Nine

The sight of a feather in a peacock's tail," Charles Darwin once wrote, "whenever I gaze at it, makes me sick!"

It was a passing thought in a letter to his friend Asa Gray, the famous botanist. Darwin wrote it in April 1860, a few months after the publication of his seminal work, *On the Origin of Species*, which explained his theory of natural selection and survival of the fittest and how slight, incremental changes allowed species to adapt and evolve over time. The challenge of explaining the peculiarities of creation with a grand theory became, at times, too much to bear. "I remember well time when the thought of an eye made me cold all over," he wrote. Yet having demonstrated quite elegantly how an eyeball would have evolved, Darwin still had to reconcile all of the nagging curiosities, the "small trifling particulars" that seemed to contradict, or at least ignore, his theory.

Like the peacock.

A peacock's profusion of unwieldy, eye-catching feathers did not fit tidily into Darwin's principles of selection for survival. Nature was full of such maddening examples. There were others Darwin could have mentioned in his letter—the call of a bullfrog, for instance, or the antlers on a moose, both of which appear to be potential disadvantages to a long and healthy life. But he went with the peacock, probably, because it's an easy visual, almost a parody of evolutionary biology: The bird's ostentatious ridiculousness, as I'd mentioned to Dean Daniel at St. John the Divine, presumably should have gotten it eaten frequently enough by jaguars to nudge it toward more subdued and shorter plumage if not hurtle it into outright extinction. The peacock's train is heavy and seemingly cumbersome, two hundred feathers almost five feet long that one would think would make running difficult and flying awkward. It is also a screaming advertisement, regularly unfolded into a glittering half-circle billboard more than nine feet across that the peacock deliberately vibrates and rattles. Survival of the fittest should not, generally speaking, favor a bird that enthusiastically announces its presence to predators it cannot easily escape.

What's more, the peacock sheds the entire ornamental rig every year, just scraps the whole train and starts over. This is, on one level, perfectly routine. All birds molt, replacing worn and damaged feathers on a regular cycle, annually for most species, more frequently for some. An adult peacock, however, is extruding feathers longer than its body, hundreds of them, at a freakish pace: To grow your hair the length of a mature peacock's train, by comparison, would take about nine years. That requires an enormous expenditure of metabolic energy that, over the eons, surely could have been diverted into a trait more useful for avoiding jaguars or finding food or anything else related to not dying.

So why would the peacock evolve to have a train—or the moose oversize antlers or the bullfrog a full-throated croak—that might endanger, rather than enhance, its survival?

Sex. Obviously.

Why Peacocks?

A peacock's display, in its entirety, is an elaborately choreographed performance. He hoists his train, which expands as it rises, falling open like a Spanish fan. Then he vibrates those feathers, and they make a sound like a roll on a muffled snare drum. He extends his wings downward and rotates the tips in a way that looks like jazz hands, and he dances in a tight, whipping arc. He does this in front of females with, as any dope can intuit, the intention that one or more of them will be impressed enough to consent to sex. He does this, in fact, on a small stage in competitive proximity to other peacocks displaying on their small stages, the whole collection of which is referred to as a *lek* but might as well be called a marketplace of very aggressive sellers. The successful male—presumably well-decorated and a gifted dancer—will then pass on his genetic material, thus propagating another generation of successful, well-decorated males, and so on. None of this is remotely subtle.

But understanding to what end a peacock uses his particular tools and how those tools would be passed on does not explain why he has those tools to begin with. A peacock didn't adapt a train in order to more efficiently crack open seeds or vigorously defend itself or make it easier to hide from predators. Under Darwin's theory of natural selection, there is simply no excuse for it.

Which led to Darwin's next theory, separate and distinct from natural selection, which was *sexual* selection.

Darwin had already begun working that out when he wrote to Asa Gray. He touched on it explicitly in *On the Origin of Species*, remarking that a colleague's peahens had been especially attracted to a certain peacock. "I can see no good reason to doubt that female birds, by selecting, during thousands of generations, the most melodious or beautiful males, according to their standard of beauty, might produce a marked effect," he wrote in that work. "[B]ut I have not space here to enter on this subject."

Twelve years after *Origins*, he published that theory fully developed in his 1871 masterwork, *The Descent of Man, and Selection in Relation to Sex*. The book was a best seller, and the second half of it asserted two radical notions: One was that creatures other than humans had an aesthetic sensibility, that they were capable of appreciating beauty; and the other was that females had the sexual agency to steer evolutionary ornamentation.

In other words, peacocks are so spectacularly accessorized, possibly to their physical detriment, because peahens prefer them that way. Exactly why has not been wholly settled; the peahens have not spoken directly on the matter. But it unavoidably suggests that the dowdy peahen, who remains drab so she can smartly hide her nest in a thicket and look after her hatchlings and do all the work, has coaxed from the male over untold generations a peculiar beauty that she finds pleasing.

That idea that the female of any species exerts such control, sexual or otherwise, landed with a thud in Victorian England. Animals weren't considered developed enough to have discerning tastes and intellectual capabilities (they are and do, as a library of subsequent science has confirmed), and it was unthinkable that females could exercise dominance over males. There were also religious objections to the whole idea of evolutionary selection, natural or sexual. The prominent art critic and philosopher John Ruskin was friendly with Darwin but also one of his fiercest critics. He did not believe that the beauty of the natural world was the result of endless adaptations, of flora and fauna struggling in an endless death match for survival—he believed all of it was created by God for people to enjoy. Beauty existed for the sake of beauty, an expression of the divine, and it needed no further justification or explanation. "Remember that the most beautiful things in the world," he once wrote, "are the most useless; peacocks and lilies for instance."

When Ruskin or Darwin wrote of peacocks, each almost certainly had in mind the species *Pavo cristatus*, the India blue. Peafowl are Galliformes, the same order as turkeys and pheasants and Comet and Snowball and other relatively large birds that feed on the ground, and the India blue—the variety of peacock with a sapphire breast and head—is the one with which most people are familiar. But there are two others.

Green peafowl, or *Pavo muticus*, of which there are three subspecies, are native to Southeast Asia and never spread in the wild beyond such steamy climes. The males tend to be taller and larger than their India blue counterparts, and they are significantly more shy around humans. Even people who raise them in captivity sometimes find them off-putting, a neurotic and fickle bird that is aggressive in the breeding season and so fragile with cold that it is, in an objective sense, the hothouse flower of domesticated peafowl.

The wild population, meanwhile, is plummeting. The International Union for Conservation of Nature put the green peacock on its red list of threatened species in 1988; as of 2009, it was categorized as endangered, two steps below extinct, with fewer than twenty thousand birds estimated to remain. Habitat loss is the main reason. Wide swaths of the green peafowls' forests and grasslands have been consumed by agriculture and housing. In China, where it's been listed as critically endangered since 2015, a planned $420 million hydroelectric dam would flood the largest surviving habitat in Yunnan province—which contains fewer than three hundred birds, or more than half the total believed to be alive in China. In 2017, a conservation group called Friends of Nature sued to stop construction; three years later, a court suspended the project until further reviews were completed, an unprecedented triumph for Chinese environmentalists and a reprieve for several hundred birds. The environmentalists' success in this case speaks to the enduring power of the peacock as a kind of universal fan favorite. Their good looks are their saving grace, with plumage mightier than the Chinese government. Magical, indeed.

The third species of peafowl is the Congo, which hardly anyone can imagine because very few people have ever seen one. They are the smallest of the peacocks, half to two-thirds scale of the other two, the males a dark violet trimmed in green, the females a subdued brown with a wash of shimmering green on their backs. They live in the Congo (the peacock naming scheme—blue, green, Congo—is refreshingly utilitarian), widely scattered in the jungle underbrush yet so rarely glimpsed by outsiders that it wasn't identified by a Western scientist until well into the twentieth century, and then only by happenstance. An American ornithologist named James P. Chapin on an expedition in 1913 noticed an unusual feather in a Congolese headdress that he recognized as a secondary flight feather, though from the wing of what bird he had no idea. Twenty-three years later, in a storeroom of the Congo Museum in Tervuren, Belgium, Chapin found two dusty, stuffed specimens that had the same curious feather. They looked a lot like peacocks. "It was a discovery of a large, major species, long after it had been assumed that no more surprises were to be expected from the Congo," a memoriam written after Chapin's death in 1964 noted. He named the bird *Afropavo congensis* and eventually found four live ones in the wild.

India blues, by comparison, are ubiquitous, often referred to as the common peacock for a reason. They are the peafowl of arboretums and English estates, of California villages and North Carolina horse farms, of ancient myths and modern art. Though they are native to the Indian subcontinent and are the national bird of India, they are global citizens of long standing: Traders and collectors began shuffling them around the planet so many centuries ago that today one is as likely to stumble upon a muster of them in a Canadian subdivision as in a forest outside of Jaipur. Depending on the translation, the Bible reports that King Solomon's fleet of ships brought to Israel "gold, silver, ivory, apes and peacocks" almost three thousand years ago. The birds were so familiar to the Greeks that Aristotle in his *History of Animals* in the fourth century B.C.E. wrote about them as domesticated game—"Peafowl live for about twenty-five years, breed about the third year, and at the same time take on their spangled plumage"—and assigned to them unflattering human traits—". . . jealous and self-conceited, as the peacock." Romans raised them as decorations for landscapes and centerpieces for feasts, and as the empire spread, so did the range of the India blue: They were in Western Europe by at least the fourth century and probably earlier. It took a while for them to cross the Atlantic, but India blues have been running wild, or semi-wild, in North America since the late eighteen hundreds.

The important thing about a peacock is his train, just as the important thing about an elephant is his trunk and the important thing about a zebra is his stripes. The train is what defines a peacock, to human eyes anyway, the thing that makes it a magnificent curiosity, an extravagant presence. It is also the part of the bird that makes us wonder how it survived past the Eocene period. And yet peacocks manage just fine in the wild.

We humans, many of us—me, at least—assumed that a bright blue bird waving an enormous and sparkling green-gold flag amid grasslands and low shrubs would pop against the backdrop. That is because we (or I) assumed that the rest of Earth's creatures see things the way we do.

They do not. Insofar as the peacock is concerned, the big cats most likely to kill it in the wild have only two types of color-receptive cones in their eyes, as opposed to three in most people and four in birds. Specifically, those predators lack red-green color discrimination, which means the shimmering plumage we see, in all probability, appears more like variations of foliage to a jaguar or a leopard. When it's limp, the spotted strands of a peacock's train most likely blend into the landscape just as well as any pheasant in the field.

The train is made up of between 150 and two hundred coverts anchored in the back of a mature male, each a long white stem lined with filaments sprouting like thin leaves. (Technically, those filaments are called barbs, but since there's nothing sharp or pointy about them and they are very soft, I'm going with a less confusing layman's term.) They come in four varieties, about three-quarters of which are topped with the iconic ocelli, though up close they resemble eyes only in a sixties-pop-art kind of way: The center is a dented oval of deep violet encircled by, in order, rings of dark blue, a coppery-bronze alloy, and finally, greenish yellow. Also, the eyespots are constructed from the same sort of filaments lining the rest of the stem, the individual strands packed so closely together that they appear to be a solid object. The longest coverts, meanwhile, end in a delicate T shape, like the tail of a dainty tropical fish; shorter ones resemble swords, curved like scimitars with blades of blue-green

barbs; and the very shortest are swords with a diminutive eyespot at the tip.

A peacock's actual tail feathers are comparatively short and wide and industrial gray and, for most of the year, hidden beneat' the prettier coverts. But it's the tail feathers that do all the work: displaying peacock is raising his tail feathers, which in turn lift / train, like a stage set pushed into position by hidden hydraulic the coverts rise, they spread out, falling to each side until the into the familiar half-circle. The long fishtails outline the cu that arc, the tips of each fishtail touching the tips of the t either side; think of a line of capital T's in a tightly spaced ro tops giving the appearance of dashes connected into an alm line. The swords form a soft, fringy edge along the bott fan, and the eyespot feathers are distributed more or l from side to side and top to bottom.

When a peacock rattles his train, the tail, again, is labor. The train appears to almost liquify, the backgrou ing like a pond stirred by a strong, sudden wind. T all of those filaments, thousands of them attached hundreds of feather stems, are swaying and waving the eyespots appear to barely move, as if they are on that rippling pond. The reason, researchers dis is that the filaments that form the ocelli are held crohooks, which allow them to act as a heavier while barbs farther down the shaft wiggle and izing illusion.

A displaying peacock's train is entrancin presumably, peahen) eye not merely becaus cause they are ever-changing. They brighte other colors altogether, turquoise to coppe

shifts his position, turns a few degrees one way, a few back the other. Ideally, he tries to angle himself forty-five degrees to the sun, an alignment that apparently offers peak gleam from a peahen's perspective.

Despite their appearance, the peacock's feathers are of no particular color at all. What look to be pretty pigments are the result of the nearly invisible structure of the feathers themselves.

Iridescence is the ability of an object to appear to change color with the angle of the light, and it has been studied in peacock feathers for centuries. Isaac Newton included them in his experiments with light in the sixteen hundreds and concluded that "their Colours arise from the thinness of the transparent parts of the Feathers." Around the same time, the scientist Robert Hooke looked at them under a new invention called the microscope. "The beauteous and vivid colours of the Feathers of this Bird," he wrote, "being found to proceed from the curious and exceeding smallness and fineness of the reflecting parts, we have here the reason given us of all those gauderies in the apparel of other Birds also." They were both basically right on the big picture: Peacock feathers are constructed to refract and reflect light into the changeable colors we see. Imagine a prism splitting a beam of white light into a rainbow—it's that general principle, only limited to a peacock-specific spectrum by its peacock-specific structure.

Three hundred years after Newton, in 2003, Chinese scientists scanned peacock feathers with an electron microscope to look at the structural details. Each of the filaments radiating from them is covered with tiny barbules; each of those barbules, in turn, is covered with a lattice of keratin—what your fingernails are made of—binding together rods of melanin—the stuff that makes skin dark. If that lattice is tweaked, if the rods are spaced

differently and such, those barbules will appear blue or yellow or some other color.

One could suggest then that the colors are *literally* coming out of the peacock, like he's taking light and manipulating it for his own delightful purposes. Which is a much better story than the one about Hera and her dead giant's eyes.

All the studies of optics and iridescence and keratin structures, fascinating as they are to people who enjoy spelunking through rabbit holes, are in a way a distraction from the elemental Darwinian point of a peacock's train, which is: Peahens like eye candy. *How* the coverts are raised or *why* the colors shift with the angle of the sun are not questions any peahen has ever pondered. It is enough that the display is beautiful.

But what does the discerning peahen look for in those fanned-out feathers? Is she impressed by the biggest peacock? Is she picking out clues among the eyespots and fishtails, deciphering which male is the healthiest, the strongest, the most virile? Or does she have a quirky taste in feathers, maybe a thing for extra swords? What, exactly, does a peahen look at—what does she *see*—when a male is displaying?

Mostly, she looks at something else entirely. For every four minutes a peacock flaunts his train, a peahen ignores him for almost three. And when she does look, she is much more interested in lower regions, in the swords and bottom-row eyespots and, from the back, the wings. Jazz hands and rattling feathers catch her attention, but she's decidedly disinterested in the grand sprawl of the show. She barely glances at the upper eyespots, which appear to be more useful as a long-distance lure poking above low bushes and high grass. Still, numerous studies suggest peahens are gathering information from a male's display and using it to make a deliberate, selective choice of mate—precisely as Darwin theorized one method of sexual selection would work.

Yet the mating ritual isn't limited to the charms, utilitarian or aesthetic, of the peacock's train. Roslyn Dakin, who has a doctorate in animal behavior and leads the Dynamic Behaviour Lab at Carleton University in Ottawa, was one of three researchers who discovered that the peahens' crest—the Seussian toodle—seems to serve a functional purpose. When a peacock rattles his train, it vibrates about twenty-five times per second. It turns out the resonant frequency of a crest, the rate at which it naturally vibrates, is also about twenty-five times per second, which suggests the crests play some role in the complex ritual of display and selection. What role, precisely, is unclear—peahens do have ears, after all, and presumably can hear the rattle just fine—but that almost certainly isn't a coincidence. Especially as it came on top of earlier research that recorded the train vibrations creating infrasonic sounds, which are too deep for humans to hear but both peahens and peacocks respond to.

Darwin nailed the big picture, but he whiffed on the soundscape. The "vibratory movement apparently serves merely to make a noise," he wrote in *The Descent of Man*, "for it can hardly add to the beauty of their plumage." Perhaps not, but it does somehow add to the overall attractiveness of the peacock to the peahen. He's trying to pass on his genes, after all, so he will employ whatever abilities nature gave him to entice a willing partner.

Like, for instance, the ability to lie. Peacocks do this with shocking frequency, which Dakin figured out when she was doing fieldwork in the Los Angeles County Arboretum, among other places.

Why Peacocks?

If a peahen finds a peacock's elaborate presentation worthy of sex, she will settle into a receptive position. From there follows a brief, almost comical copulation: The successful male makes a short, swooping run at the female and lets out a boastful hoot, almost like the sound of a drawn-out clown-car horn. Hoot-dash, scientists call it (though hoot-scoot would be much more fun). Other peahens will hear that sound, assume the peacock making it has been vetted by other hens, and thus consider mating with him, too. It's the "I'll have what she's having" theory of peafowl reproduction.

One third of those hoots are fake. Really. Dakin counted. She watched and she listened and she counted, and for every three hoots, there were only two scoots.

That is the most peacocky fact ever discovered about peacocks. As if the trains and the rattling and the dancing aren't desperate enough, those big birds routinely misrepresent their sexual prowess to try to impress peahens. It's astonishing. Since at least Aristotle's time, humans have been accusing peacocks of embodying unpleasant human foibles, pride and vanity, and yet somehow missed the most obvious one of all.

Chapter Ten

In early February, I found Carl lying in a clump of hay. He didn't get up when I sat on the cinder block, and he didn't move when I tossed him a blueberry. I took a few steps toward him, and he roused himself only enough to shuffle farther away. He wasn't bleeding that I could see, so I guessed he hadn't lost a fight with Mr. Pickle. He settled into another patch of straw and looked at the wall.

I remembered Dr. Burkett saying birds are very good at hiding illness, so if one of them looked sick, he was really sick.

It was almost three o'clock on a Friday, and I had to pick up Emmett from school. On the way, I called Burkett's office and told Julie I had a sick bird and that I was going to bring some droppings to the office. I assumed Burkett would want droppings, as feces seem to be a standard multispecies diagnostic medium.

Julie said Burkett would call me when he was finished with a patient, but I told her not to bother. I'd be there as soon as I could.

Burkett was standing next to Julie's desk when I opened the door. "You remember me?"

"Oh, yeah," he said. "The guy with the peacocks."

"Excellent. One of them is sick. I brought poop."

He laughed. "That's what I like, poop on a Friday afternoon."

I'd collected three specimens in separate baggies, hoping I'd gotten lucky and one of them was Carl's. Burkett took all three back to his lab. I sat down to wait but instantly got antsy, got up, looked around.

"How's Elvis?"

Julie nodded toward the sign on the wall. She hadn't been bitten in nine days.

Elvis chattered at me, pulling himself along the side of his cage with his beak. In the cage behind his, which I hadn't noticed before, was a big gray parrot. It was sitting on a wide wooden platform instead of a round roost, like a dowel rod. I peered through the wire. The bird had no toes, just stumps at the end of its legs.

"What happened to him?" I asked.

"Her," Julie said. "Nothing. She hatched that way."

"So it's a birth defect?"

"Yeah. Her parents were old."

There was a pink bird in a cage against the far wall. "Adopt me!" read a hand-lettered note attached to the frame. The sign said her name was Rosie, and she was a four-year-old rose-breasted cockatoo. I asked if her owner had died.

"Rosie? No, they're just giving her up."

"Why would someone go to all the trouble of getting a bright pink bird and then give it up?"

"She ate something," Julie said. "I don't know what it was, but it won't pass, and she needs twelve hundred dollars' worth of surgery. So they're hoping someone will adopt her."

"Oof. That's a high-dollar bird."

I leafed through a bird magazine. A woman came in with her son, a boy about Emmett's age, and a noisy rooster in a plastic bin.

Burkett popped his head out, motioned for me to wait, turned to the woman with the rooster. "Come on back," he said to her.

Another thirty minutes passed. The woman and her son left with an empty bin. I could hear the rooster squawking in the back, an inpatient now. Then Burkett came out.

"Well, good news, no parasites," he said. "But in one of the samples, probably Carl's, it's got some budding yeast in it."

"Yeast? And that means?"

Burkett drew in a breath, puffed it out. "All I can tell you right now," he said, "is that it means his immune system isn't working right. We'll give him a full workup, see what's going on."

"Okay. How do we do that?" I asked.

Burkett hesitated, unsure of the question. It was, I had to admit, an odd use of the royal *we*, considering I wouldn't be doing any of the working up. "Well, you bring him in, and I take X-rays and draw blood—"

"See, that part right there, the 'bring him in' part. How do we do that?"

"You catch him. Put him in something, like a feed bag, and bring him here. You have a feed bag?"

"Yeah, yeah, I've got that. I've just never done the catching."

Surely I couldn't be the first person who didn't know how to catch a peacock. Burkett smiled, nodded in recognition. He explained his preferred technique, which was to wait until dark, when Carl was roosting and couldn't see as well, then sneak up from behind on a ladder. In a single, swift movement, I was supposed to get one arm around his body to pin his wings and, with my free hand, clamp his legs. Then I would back down the ladder in the dark with both arms occupied by a large, unhappy bird.

That hung between us for a few seconds.

"What are the odds of me getting hurt in all this?"

"You'll be fine," he said. "As long as you're fast. But, like, milliseconds fast."

"Uh-huh. And if I'm not?"

"Oh. Well, pretty good, I guess. Black eye, probably. They can really get those wings swinging. Maybe a broken nose."

"What about the, you know, what do you call them, the talons?"

Burkett drew in a breath and pulled his shoulders back, as if it hadn't occurred to him that I might not be a gifted leg-grabber. "Yeah, those," he said. "If he catches you with one of those, you're gonna bleed." Pause. "If he catches you good, you're gonna bleed a lot."

I nodded. I assumed I would bleed a lot. "How about a net?"

He seemed to be considering this for the first time in his career. "Yeah," he said. "A net would work. I don't see why not."

Problem was, I didn't have a net and wasn't sure where to find one at six o'clock on a Friday evening. Maybe a sporting goods store. Or a fisherman. Did any of my friends fish? Uncle John! We call him Uncle John because he's Emmett's godfather. He fished. He fermented his own turmeric and roasted his own coffee beans and smoked anything that would fit in the barrel-shaped cooker on the patio, which was beside the point, but it meant he had a lot of useful accessories. He would have a net. Probably one of those big long-handled things, too, like for hauling in salmon or some other fish about the size of a peacock.

I'd met John when we both moved to Durham, but I'd known his mother and father for years before that. John's parents were Louise's parents' closest friends. Her father and his father were fishing buddies and hunting buddies and drinking buddies. John's father was an Episcopal priest who married me and Louise, both of her sisters—one of them twice—and both of her parents, though not to each other. He baptized Calvin and Emmett, with those last two events accounting for the entirety of my church attendance in the last half of the aughts, a fact upon which John's father never commented, hinted at, or alluded to. He also poured very dry and generous martinis. He was my kind of preacher.

I called John from the parking lot of the Birdie Boutique and explained my dilemma. He didn't have a fishing net, but he was pretty sure he had something in the basement. He kept me on the phone while he clomped down the stairs and rummaged around. "Got it," he said. "Yeah, this should do it. Just let me get it off of here and I'll bring it over."

I told him I'd pick it up, that I didn't want to interrupt his Friday night any more than I already had.

"I don't mind. I want to see. Wait—do you want help?"

"Oh, sweet Jesus, yes. Really?" To my initial surprise, our friends had shown no more than a passing interest in the peacocks. People would come over for drinks or dinner and stand outside the garbage coop for a few minutes, waiting for Mr. Pickle to do his peacock thing and throw up those feathers. He did not perform on command, however, and I hadn't accounted for the fact that looking at someone else's caged peacocks gets boring pretty quickly. It's like sitting through a violin recital of eight-year-olds: Sure, it's cute and all, but once your kid is done, the afternoon tends to drag a bit. "I'm a little scared someone's gonna get hurt," I told John. "Me or the bird. Probably me."

"Yeah, man, I'm on my way." He was half laughing. "I've been telling the girls you should always go help people if they ask and if you can. Helping you catch a peacock is an excellent example."

•   •   •

There was no handle on John's net. He had it wadded up in his hand, a tangle of yellow nylon that he shook out with a couple quick snaps of his wrist. It opened into a square about four feet on each side, thin strings knotted together in a two-inch grid.

"This gonna work?" John asked.

"It's gonna have to, I guess." I'd hoped John would bring something I could use from a distance, not get within slashing range until Carl was contained. But this was a close-quarter net, the kind we'd have to drop on him. "Where'd you get that sad thing?"

"I got it out of a soccer goal I made Julia."

"You cut up your daughter's stuff to help me catch a sick bird? I'm touched. Yeah, we'll make it work."

Ethel and Mr. Pickle scooted to the far side of the pen when we stepped in, but Carl stayed in his straw pile. We opened the net, John holding one end, me the other, and advanced on him slowly, trying not to spook him into bouncing off the chicken wire. He let us get within three feet before he stood up and backed away. We were probably just prolonging the whole thing: Carl moved at our pace, keeping a steady distance between himself and the net until we pushed him into the corner.

Then he bounced off the chicken wire.

It was a gentle bounce. Carl didn't make his move until he was a step away from the edge of the pen. He couldn't get any speed, and there was nowhere for him to go. We tossed the net over him, and I dropped to my knees and smothered him like a fumbled football. He flapped a wing free and I grabbed it, tucked it against his side. He was either too tired or too resigned to struggle after that. He was also remarkably soft, his chest almost silken. I'd never touched any of the peacocks before, except to unload them, and that was just feet and feed bags.

"Well, that was easier than expected," I said.

I looked up over my shoulder at John. He had this expression, a look particular to him, a sort of awed smile just on the edge of laughter that suggested he'd discovered a very cool thing and would very much like to do, consume, or observe more of it. It was a look of reckless, confident optimism.

"All right, man," he said. "Now what?"

Once I had a grip on Carl, John zip-tied his legs, then cut a hole in the top of an empty bag of Purina Game Bird Chow. I figured he was going to be in the sack for a while, maybe overnight, and would appreciate being able to stick his head out and look around. It didn't occur to me that Danielle hadn't mutilated her sacks for a reason: Once we maneuvered Carl into the bag and his head through the hole, he started wiggling, forcing his way out. The hole got too big, so John dumped a quarter-sack of Mule City chicken feed in Comet and Snowball's pen so we could start over, only without the hole.

We got Carl cinched into the Mule City bag, and then I realized I had no idea what to do next. Leaving him in a sack on the ground for the night wasn't ideal, but neither was taking him into the house. I hadn't thought this through. It was almost seven-thirty, ninety minutes past closing time, but I called the Birdie Boutique anyway. Maybe the phone would forward to Burkett's cell, I thought, and he could tell me what one does with a bagged-up bird until morning.

Julie answered on the first ring. She has a voice engineered for the telephone, soothing and steady. Burkett was still there, she told me, and it would be best if we brought Carl in right away, let him get settled there so Burkett could start on him first thing the next day.

I dropped off Carl not long after. Burkett took him to the back. "Water only," I heard him tell an assistant. He returned to the front desk, asked me to bring food in the morning, and said he'd try to have some answers by nine.

Julie was unlocking the front door when I pulled in the next morning with a gallon baggie of bird chow. "Am I early?" I asked as I climbed the porch stairs.

"Oh, no, he's been here for hours," she said. "Just because we're not open doesn't mean he's not here."

I followed her in and could hear Burkett moving around in the back. I waited only a minute or two before he popped his head out and waved me into the room where I'd first met him in July. "I found a lot of the problem," he said, clicking a mouse a couple of times until an X-ray lit up the computer screen.

It was clearly an image of a bird on either his back or his stomach, legs splayed, wings outstretched, the same pose as the eagle on the back of a dollar bill. "All these white spots," Burkett said, swirling the cursor around what appeared to be Carl's stomach, "they shouldn't be there. Those are all little pieces of metal."

"*Metal?* What kind of metal?"

"Not sure. But look at this." A new image popped up, a negative of an X-ray, the metal showing up dark against lighter soft tissues. There was a large collection of black dots and wedges and shards and, just left of center, a round hoop, like a thin donut. As we were looking at that, a printer started unspooling the results of Carl's blood work, a long register receipt of abbreviations and numbers. "See that one?" Burkett said, pointing at a 2,400. "That should be three hundred. That means his kidneys aren't working."

This was escalating at an unfortunate pace. The day before, I'd assumed he had worms, not multiple organ failure.

Burkett scanned the rest of the numbers. My adolescent bird, he told me, had lead poisoning and probably zinc poisoning. I remembered raking the pen before I'd brought the birds home, pretty sure that I'd cleaned up all the dangerous fragments. I must have overlooked something. Or maybe those were misfired staples I might have left scattered when I was putting up chicken wire. "He's going to need his blood chelated to avoid neurological damage," Burkett said.

"Like Keith Richards!" I said, a bit of trivia that brought momentary joy. It was mostly urban legend, but when I was a kid, everyone knew that the Rolling Stones guitarist had his blood pumped out and cleaned and put back in so he could get off heroin, which always struck me as a very rock-star thing to do. The story always got shortened to "Keith got his blood chelated," though that's not even what the word means.

Burkett gave my joke an obligatory smile. Carl, he explained, would get dosed with intramuscular drugs that would bind to the metals in his bloodstream so they could be flushed out with his urine. "We'll give him a couple of days with that, and on Monday I'll go into his stomach through his crop and pull out as much of that stuff as I can and try to flush out the rest. If that doesn't work . . . well, things could get complicated."

I rubbed the back of my neck. "What's the prognosis?"

"Oh, a hundred percent," Burkett said. "I can fix him."

I felt my shoulders slump, an involuntary deflation that surprised me. I appreciated his confidence, but damn, those were expensive words. If he'd told me Carl was going to suffer horribly and probably die anyway, we could euthanize him with a minimum of guilt. Killing him would be almost noble, a sad yet humane end to his misery that the doctor and I would stoically execute. I didn't *want* Carl to die, but veterinary therapies that get complicated are never cheap.

Carl was just a bird, I told myself. I didn't even like birds a year ago. And was neurological damage really so bad? Maybe it would be minor. He'd always been a skittish little dimwit. And he was barely a proper peacock, what with that mangy train and one silly, cockeyed eyespot.

"Can I see him?"

"Yeah, of course." Burkett pointed through the operating room toward a small room stacked with cages. Carl was in one on the floor, the biggest box there, but even his stubby coverts had to be curved to fit. The stalk of his one eyespot had gotten creased halfway up, and it bent back to hang limply near his thigh. He seemed to be staring at the blank back of the crate.

I squatted next to the cage. "Hey, Carl," I whispered. "How you doing, buddy?" He twisted his head slightly toward me, blinked twice, then turned back to the wall. His toes were stretched over the holes of a plastic grate that allowed his droppings to fall through to a tray. There wasn't enough room for him to turn around, let alone walk. I doubted he could sit. He looked miserable, had to be miserable, all cramped and poisoned, and I didn't believe that was my presumptuous mind reading. I took a deep breath, held it, let it out through my mouth, a calming technique a shrink taught me once. Stimulates the vagus nerve.

"All right, Doc," I said loudly enough for him to hear me two rooms away. "Do what you gotta do."

Chapter Eleven

Peacocks for centuries had been a luxury of the wealthy, at least in the Western world.

In their native East, where India blues are abundant in the wild, almost as common as macaques or geckos, they were a bird of the people. A companion to gods, yes, shaded with mysticism, and a symbol, depending on the circumstances, of prosperity and beauty and wisdom. But the peacock's charms were not reserved for the rich. It was a common bird for the common man.

Once they were caged up and hauled away by traders, however, peacocks became objects to own, to covet. The mythical stories grew darker—remember Hera and her dead giant—and the bird came to be more explicitly associated with a kind of ostentatious wealth. A peacock was to be boastfully displayed, whether in the courtyard or on the banquet table. Henry III, for example, had 120 of them served at his Christmas feast in 1251, and the Archbishop of York had 104 prepared for a feast in the fifteenth century. Yet the peacock's beauty was always the point, even when it was being eaten: A regally prepared peacock would be skinned and dressed, roasted, and then covered again with its own skin for serving, the feathers still lovely and gleaming. The menu for a relatively small banquet for France's Lord of Foyes in the fourteen hundreds—three courses followed by fruit—included a second course of "stuffed kid shoulders, sea pullets, young peacocks in full display, quails with sugar."

The flavor and texture of the meat wasn't the primary appeal. "All present at [a medieval banquet] would admire the beautiful bird when it was brought ceremoniously to the table of the host," a Dutch food historian and onetime peacock chef named Christianne Muusers wrote. "The admiration was probably somewhat less when it came to eating the peacock: its meat is very dry. No wonder the peacock disappeared from the tables when the turkey . . . made its entrance in Europe during the sixteenth century!"

It's unclear when the first peacocks arrived in America, or who brought them, but it was no later than 1870, when the landscape painter Frederic Church had them scratching around his Hudson Valley estate. There is no doubt, however, that they were imported strictly as ornaments, as there was no practical reason to bother. Peacocks were not viable game birds, as they are too big, too slow, and too easy to spot to make much sport of shooting them (though some people do anyway). Nor were they worth raising commercially for food. Some modern chefs and home cooks have had success preparing them—a long simmer in a Crock-Pot reportedly helps—but they grow more slowly than chickens, take up more space than turkeys, and taste worse than both. India blues, ever alert and exceedingly vocal, do make for fine watchdogs, but so do other animals, like, say, dogs.

That is all to the good of the peacock. Its most significant feature was not designed, by the gods or by nature, to be anything other than lovely; that chickens and turkeys are plainer and tastier is to the great detriment of their miserably short lives. The peacock's unsubtle beauty, by contrast, often made it part of a show, a set piece in the act of being a certain kind of rich at the beginning of the Gilded Age. (Peahens never really figured into this other than as incubators for more peacocks.) Even the word itself, *peacock,* was evolving into useful shorthand for a type of showboating excess.

In the early sixteen hundreds, for instance, Shah Jahan, perhaps the greatest of the Mughal emperors and the one who built the Taj Mahal, commissioned an elaborate throne for himself. It took seven years for the finest goldsmiths and artisans to create a platform roughly six feet by four feet set on four golden legs, from which a dozen columns rose to support a silk canopy. More than a ton of gold was hammered and shaped to form the structure, and it was inlaid with more than five hundred pounds of gemstones, including 116 emeralds, 108 rubies, and more diamonds than anyone bothered counting. The twelve columns, according to a French jeweler who saw the throne before the Persians looted it in 1739, were each covered with rows of pearls. The throne also had several showstopper gems, such as the Koh-i-Noor, a diamond the size of a ping-pong ball whose name means "mountain of light." It is reputed to be the most expensive throne ever crafted; one estimate suggests it would have cost $1.2 billion in today's dollars, and it was widely claimed to have cost twice as much as the Taj Mahal, which would peg the number at $1.8 billion or so. The Mughals called this sparkling perch the Jeweled Throne or the Ornamented Throne, perfectly appropriate titles for a canopied seat encrusted with ornamental jewels. Decades later, however, after the throne had been destroyed and the gems scattered—the Koh-i-Noor ended up in the British crown jewels, where it was cut down from 186 carats to just under 106—Western historians started calling it the

Peacock Throne. That name is not *inaccurate*; peacocks were an important symbol in Mughal design, and they were incorporated into the throne, though accounts of exactly how differ. But amid the rubies and the gold and the pearls, birds were hardly the most notable feature. Calling the throne a peacock just encapsulated the gaudy spectacle more clearly.

The birds in Frederic Church's day still had an air of exclusivity about them. Live ones were uncommon in North America in the late eighteen hundreds, but stuffed ones were advertised to the well heeled as household decorations, notably as screens for sooty, unlit fireplaces. Over the next century, peacocks would go mass market, selling everything from cigarettes in the nineteen teens to condoms in the thirties and forties to Jell-O in the fifties. The bird went thoroughly mainstream in May 1956, when NBC introduced its peacock logo to highlight the increasing number of color broadcasts and, more to the point, to sell color TVs for its parent company, RCA. That was an appropriate position for the peacock, fronting for the modern medium of fables and mythology. Twenty years later, the bird either fell more deeply into mythology or it collapsed completely into kitsch, depending on one's opinion of Vegas-era Elvis: His favorite costume of 1974 was the peacock jumpsuit, ten thousand dollars of white fabric and brocade and gold lamé, a stylized bird across the chest, plumage raining down the right leg, and another on the back with the train sweeping down the left leg. He wore it on the cover of his *Promised Land* album, and a collector bought it in 2014, sweat-stained and with a broken zipper, for $245,000.

Around the same time India blues were arriving in the United

States, the peacock was making a resurgence as a subject of serious art, as opposed to the commercial silliness to follow. For centuries, the bird had been incorporated into paintings and etchings and mosaics, its popularity waxing and waning. By the late eighteen hundreds, it was at an apex, "the go-to image of the 19th-century's Aesthetic Movement," wrote Michael Botwinick, the former director of the Hudson River Museum, which in 2014 hosted an exhibit called *Strut: The Peacock and Beauty in Art.*

Perhaps the most famous peacock art to emerge from that movement was a piece by James McNeill Whistler. In 1876, he was living in London and doing some work in the Kensington home of a shipping magnate and art collector named Frederick Leyland. Also working in the house was an architect, Thomas Jeckyll, whom Leyland had hired to design a dining room to display his collection of Chinese porcelains. Jeckyll asked Whistler for advice on a color scheme to best show off the intricate blue patterns on the white porcelains. Yellow, he suggested. And then Leyland left the city on business, and Jeckyll took ill and stopped overseeing his project, and Whistler went hog wild with the yellow. Gold, actually, and blue, though not a bright peacock blue. He painted everything that summer: the antique leather wall panels and the wainscoting, the cornices. The motif, daubed onto the walls and ceiling, was of semicircles mimicking the green-gold scales at the base of Mr. Pickle's neck. Pairs of peacocks covered the folding shutters, and Whistler painted the walnut shelving meant to display Leyland's porcelains. Whistler titled it *Harmony in Blue and Gold.*

It's still around, the room restored and fully assembled in the Freer Gallery of Art, which is part of the Smithsonian Institution in Washington, DC. Louise and I took the boys to see it one rainy spring day, though really, for them, it was more of a soggy forced march than an adventure. Adolescent boys rarely appreciate art museums to begin with, and they tend to be even less enthusiastic when Dad is dragging them along for work.

Whistler, of course, was a genius, and the room is a masterpiece. Not the sort of thing I'd want in my house; I imagine it would run more toward the taste of a nineteenth-century industrialist. The room is roughly twenty feet wide and thirty feet long, and the shelves were lined with porcelains, just as Leyland had intended. The shutters were closed—the museum opens them only on the third Thursday of every month to prevent the work from fading in the sunlight—so the peacocks painted in gold were on full display. At one end of the room was a Whistler portrait, *The Princess from the Land of Porcelain,* which Leyland also owned, as he was Whistler's patron at the time.

I stood in the center of the room, made a slow pirouette, and did a visual sweep of the walls. "What do you think?" I whispered to Calvin. I was pretty sure I didn't have to whisper, but my voice instinctively drops in museums and churches.

"It's cool, I guess," he said. "Can we go now?"

I gave him a quiet half-laugh. We'd met one of Louise's sisters and her kids in DC, and I knew the boys would rather be running around with their cousins. "Yeah, go on," I said. "I'm gonna stick around a little longer."

The mural on the south wall of the room, the one Leyland would have seen when he sat down to dinner, was my favorite. It was the last part Whistler painted, and it told, in one broad panel, the early history of *Harmony in Blue and Gold.*

Leyland was out of town in the summer of 1876, when Whistler began painting his dining room. Upon his return in the autumn, Leyland was wildly displeased with the redecorating (upon further consideration, maybe the shipping magnate and I weren't so different after all). He was even less pleased with Whistler's invoice for roughly two hundred thousand dollars. There was an acrimonious fallout, and while Leyland eventually agreed to pay half the fee, the two men never reconciled. (Jeckyll, meanwhile, never recovered at all: After his efforts were covered in paint, albeit by a master, he had a mental breakdown in 1877 and died in an institution four years later, at the age of fifty-three.) But when Leyland again left town, Whistler slipped into the house to paint the leather panel spanning the south wall of the room. He created a mural of two peacocks, one passionate, animated, rising in defensive anger, his silver toodle echoing the white in Whistler's hair. The other stood with his wings spread, imperious, silver shillings scattered on the ground beneath him; the feathers on his chest were round, like coins, a bird decorated with crass money. Whistler called it *Art and Money; or, the Story of the Room.*

Leyland, perhaps sensing the value of an entire room painted by Whistler, did not have it dismantled or repainted. The Peacock Room, as it came to be called, remained intact, if tarnished by cigar smoke, until Leyland died in 1892. As it happened, the room was built from panels that could be disassembled, so the subsequent owner of Leyland's house was able to sell it. Charles Lang Freer, a Detroit art collector who made his fortune in railcars, bought it in 1904. He bequeathed his Whistler collection to the Smithsonian, which is how it eventually came to be fully assembled in the Freer Gallery of Art after he died in 1919. The boys weren't impressed, but Louise and I lingered, examining porcelains but mostly studying the birds. They looked nothing like Carl and Mr. Pickle, but I suspected that was because I didn't imbue them with the same passion that Whistler did the ones he put on those walls.

Chapter Twelve

Monday morning, Burkett was in blue scrubs, wide-eyed by the side of Julie's desk. "Man," he said, slowly shaking his head, "this just keeps getting worse."

"It's been two days," I said. "How much more could go wrong?"

"It's all the stuff that was already wrong," he said, turning toward the back and motioning for me to follow him. "I'm still finding things out."

Carl was on the operating table, standing up, eyes closed and wrapped in blankets, a bandage taped to his neck about halfway down. The floor was littered with bloody swabs and torn packaging of medical supplies. Burkett had been pulling rubble out of my bird for hours, all of which he'd collected in stainless-steel bowls. There were pebbles and sparkles of glass and undigested feed and bits of his stomach lining that looked like soggy rice paper. There was also that donut we saw in the X-ray. It was a copper grommet, almost the size of my fingertip.

"He's gotta have copper poisoning, too," Burkett said.

"Bad?"

"Really bad. Very toxic. Don't see it often, though. I had a chicken come in a few months ago, got it from a pebble with a little speck of copper ore in it."

Carl had a whole grommet in his gut.

Burkett's theory was that Carl had scratched the grommet out of the dirt—it could have been there from an old tarp that rotted away fifty years ago—got a stomachache, and ate all the pebbles and other rubble to make himself feel better, except he only made himself sicker because some of that rubble was poisonous. There was, oddly, some relief in that. After Saturday's diagnosis of lead poisoning, Louise had an acute panic that Carl had picked it up from the dust around the barn. Considering the boys used to run around in that dust, and still played hockey and basketball in the old hayloft, she was already researching lead-abatement companies. Because the idea that you might have done permanent neurological damage to your children is the kind of anxious thought that has to be promptly addressed, I deployed the most sophisticated diagnostic tool I could find on a Saturday afternoon: a couple of store-bought lead-test kits, the ones that tell you if the ancient paint on the windowsill is toxic. I rubbed fiber swabs in a dozen different spots around the garbage coop that afternoon. The results were all negative, which was encouraging but, I assumed, not scientifically reliable.

If Carl's primary problem was a chunk of copper in his belly, that was good news. If he'd given himself lead poisoning from eating pebbles flecked with lead and, most likely, fragments of actual lead, instead of breathing in ambient dust, then Louise could stop pricing out decontamination procedures. I was making some optimistic leaps, but none of the other birds was sick, and the pediatrician had never raised concerns. The toxins, it seemed, were contained to Carl.

I stroked the back of his neck and he opened his eyes halfway, then shut them again. He was groggy from the anesthesia. One of the two nurses in the room pulled the blankets tighter around him, lifted, took Carl back to his cramped cage.

"I got a lot out, but there's still some pieces in his intestinal tract," Burkett said. "I'm hoping to flush them out with Metamucil. Works just like with people." He picked up one of the stainless-steel bowls, swirled it. The pebbles plinked against the sides, a hard, cold sound. "We'll give it a day, take some X-rays, see where we're at."

"Is he suffering?" I asked. "Is he in pain? Is this prolonging something awful?"

"Oh, no, he's fine," Burkett said, flicking at something in the dish. "I mean, not *now*, because he's waking up from surgery. But other than that."

"And what's the prognosis now?"

"Hundred percent. I can fix him. First I want to get all this fluid out—we put a lot of fluid in him—and then see what it looks like."

I could see Carl through the doorway, bloated and woozy in his cage.

"Is he just getting better to come home for an ass-kicking from Mr. Pickle?" I was sure no one had ever asked that question of anyone, ever, in any context. "You know, with mating season and all?"

"There'll probably be some reintroduction issues," Burkett said. "They can be pretty hard on a bird coming back into the flock."

"After a week?"

"Sure. Chickens can be, like, three days. But that's all right. I'll walk you through it."

That was somewhat comforting, his certainty that Carl wasn't in agony and would fully recover. But only somewhat.

"I gotta be crass," I said. "How much am I in for?"

He seemed caught off guard. The type of people who get surgery for their birds probably aren't the same people who inquire as to cost. "Um . . . a lot. I mean, we were in there for hours this morning—"

"Doc, you don't have to justify it, really. I'm not looking for an accounting, just a number."

He looked up at the ceiling, down at the floor, up again, adding in his head. He gave me a number.

"Oh, fuck me," I said. "Seriously? Shit."

Neither of us said anything else.

Carl was not magical. He had not sprung from a dewdrop. He could not possibly be trusted to guard the gates of Paradise, and he would not find favor with a goddess. He was a clumsy adolescent who lived in a death pit built from garbage by a selfish person who wanted a novelty pet.

In through the nose, out through the mouth.

"But we're through the expensive part, right?"

"Well, I want to see how the Metamucil works first. But I think so."

I was already tired, distracted, too, by the time we sat down to dinner that night. Between trips to the Boutique, panicked lead swabbing, Calvin's hockey game, and a deadline, I hadn't caught the boys, or even Louise, up on Carl completely.

When we sat down at the table, Calvin cut right to it. "Is Carl going to die?" That caught Emmett's attention. He looked stricken. Any pet is more loved when it's in peril.

"No, no," I said. "He'll be fine." I told them about the grommet and the pebbles and the Metamucil. Louise, who'd been a health reporter for a time, told the apocryphal Keith Richards story—"So Carl is like a rock star!"—but it failed to lighten the mood.

Why Peacocks?

"I promise you, Carl's not going to die," I said. "Dr. Burkett can fix him. It just might take a while. And cost a ridiculous amount of money."

I knew I shouldn't have said that even as my tongue and teeth were forming the last syllable in *money*. The boys didn't need to know that. Cost wasn't relevant right then. I glanced sheepishly at Louise. She shifted her eyebrows enough to ask how much *ridiculous* was, and I shifted mine to suggest we should talk about that later, and then we both smiled at the boys.

"Is Carl too expensive to fix?" Emmett asked. Children are drawn to the unspoken words, like hawks to field mice.

I popped a forkful of broccoli into my mouth before I answered. A stall tactic. His question was blunt, but there was nuance in the implications. By any standard I would have applied eight months earlier, yes, Carl was too expensive to fix. That is, if the fates had whispered then that medical care for a bird that lived in a cage in the yard would cost this particular amount of money, there would not be birds in the yard. But last year's standards were irrelevant. We now owned three peacocks, two boys and a girl, and we were responsible for their food and shelter and health care, even if it involved chelation and surgery. Could it get too expensive? Was there a hypothetical point where the sunk costs would be worth writing off? If Burkett told me the price going forward would double or triple, would that make Carl too expensive to fix? Not philosophically, just as a matter of financial practicality.

I chewed very slowly, considering my options. *Maybe* was a terrible answer, the kind that leads to nightmares and therapy. I remembered the exotic animal vet didn't charge me for telling Emmett his snake was dead.

"No, not at all," I said. I was locked in now.

Emmett kept looking at me as if that answer required elaboration.

"Do you remember Otis?" I said.

"The old orange cat?"

"Yeah, the old orange cat." That was their first pet, but he'd been dead for a few years. "I got him when he was a kitten, runt of a litter that lived in a vacant lot across the street. When he was about eight or nine, back when we lived in Boston, he got really sick. He had a hole in his diaphragm, this muscle between your stomach and your chest. Probably had it his whole life. And then one day all of the organs in his abdomen, his stomach and his kidneys and his liver, popped through that hole into his chest."

Emmett widened his eyes in alarm. "Ouch. That must've hurt."

"Yeah, I'd think so. The vet told us we should put him to sleep because the only other option was this surgery that was stupid expensive. More than Carl. They'd have to split him open, put all his organs back where they belonged, fix his diaphragm, staple him shut. And then we'd have to feed him by hand and clean up after him until he got better."

"What'd you do?"

I smiled, tipped my head toward him. "Pup, that was five years before you were born. Yes, we got the surgery. Your mom—who is wildly allergic to cats, by the way—fed him pureed chicken four times a day for two weeks, and he lived another ten years."

Emmett relaxed a bit, visibly relieved.

"Sorry," I said. "That wasn't supposed to be a cliffhanger story. The point is, Carl's not suffering, and Dr. Burkett can make him better."

I should have stopped right there. But my brain continued emptying through my mouth. "Now, if he was going to die no matter what or he was in terrible pain, then we'd think about it differently," I went on. "But it's not Carl's fault he's sick. Well, actually, it is sort of Carl's fault, but I guess he didn't know he shouldn't eat copper grommets. Either way, he's our responsibility. We decided to have these birds, so we have to take care of them. It's no different than Tater."

"What about Comet and Snowball?" Emmett asked.

Louise was giving me side-eye. She knew I was making the world far too complicated for a fourth-grader worried about his pets.

"Well, chickens are different—" I stopped myself. Why were the chickens different? If I were a chicken farmer, there would be a cost-benefit question. But I'm not. We have pet chickens. I trained them to perform a jumping trick, like circus chickens. Those birds come when they're called. They recognize our voices. They express joy and confusion and something like affection. How are Comet and Snowball, in that sense, any different than Tater, let alone Carl? Where is the line between animal and pet, and did I really want to try to draw it right then?

"You know what," I said, "we're just gonna hope the chickens don't eat any copper."

Chapter Thirteen

When Kristina Frandson was a little girl in Idaho, back in the early aughts, her great-grandmother wanted to be her pen pal, which was a thing that people who grew up before the invention of email continued to do out of habit and tradition. She would mail Kristina letters from California, where she lived in a white ranch house at the top of a shallow canyon, in the hope that Kristina would reply in kind and they would continue a proper correspondence.

Kristina typically did not. Being a child, she was an unfortunate choice for a reliable pen pal. She loved getting the letters, though, because her great-grandmother wrote in a flowing cursive on hand-made stationery and, more often than not, folded the paper around small iridescent feathers, sparkling like jewels, from the train of a peacock. She was able to do that, Kristina knew, because she lived in a wondrous place where there were real, living peacocks in the yard. Kristina had seen them many times because she visited her great-grandmother, who lived with Kristina's grandmother and grandfather, with some frequency. The ranch house was on the Palos Verdes Peninsula south of Los Angeles, three lots from the dead end of a street called Dapplegray Lane. Two other lanes, Buckskin and Sorrel, branch off of Dapplegray, which is why the neighbor-hood locally is known as the Lanes. The place is only twenty-five miles from downtown L.A., but the physical distance is a pittance relative to the atmospheric remove. Much of the peninsula is semi-rural, as is the Lanes, 168 houses, mostly modest and midcentury, on large lots sloping into canyons. There are no sidewalks or street-lights in the Lanes, but the neighborhood is surrounded by bridle paths that connect with miles of other trails in Rolling Hills Estates, and there is an equestrian ring at the bottom of a track that runs down from Buckskin. It all feels secluded without being isolated because Dapplegray is the only way in, and each of the three streets dead-ends, Sorrel to the east and Buckskin and Dapplegray at the northern edge, beyond which is a nature preserve that is a haven for an indigenous butterfly, the Palos Verdes blue, that is struggling to keep from going extinct.

There used to be many dozens of peacocks and peahens roosting in the pepper trees and scratching at the slopes and slow-walking the streets, and they had been there forever, or at least since long before anyone thought to pave three dead-end lanes off of Palos Verdes Drive. Kristina's great-grandmother used to feed them in the backyard, bread mostly, an indulgence that peeved her daughter, Kristina's grandmother, no end. Kathy Gliksman, the grandmother, liked the birds just fine, but they'd leave droppings on the deck. "It isn't that you can smell it," she says. "It's just that it's poop and, you know, *there it is*. Dear God, if you leave it sitting in the sun, it takes days to dry out enough where you can knock it down with a broom."

One particular peacock had a twisted foot, the right one, gnarled from an old fracture that didn't heal right. His toes didn't work as well as the other birds'—he couldn't hold the branches in the pep-per trees as tightly—so if the wind got to blowing hard, he'd settle in on the Gliksmans' deck and lean against the wall of the house. He was not at all a pet, but he was familiar, as if he'd chosen their place for sanctuary. Also, he was the only peacock a little girl from Idaho could pick out among all the others, on account of his mangled foot. Kristina named him Beauty, not because he was especially beautiful but because that is not an inappropriate name for any peacock.

Beauty gradually got comfortable enough with Kristina that he ate out of her hand. She was six or seven years old, so she was a little intimidated by the size of him, a sharp-beaked giant with one good set of talons. But he was gentle, never aggressive, this big, wild bird taking treats from a child. "And that adds to that whole magical thing," Kristina told me many years later, "you know, that they're these wild animals and you're making that connection with them." She might not have been an ideal pen pal for her great-grandmother, but they bonded over the birds and especially over Beauty.

Beauty died on May 3, 2014. He was found at the bottom of the deck stairs in the Gliksmans' backyard. A veterinarian determined that he died probably because he was very thin. But Beauty also had buckshot in his breast, and lodged under his skin, that would have slowly poisoned him with lead, the least unpleasant symptoms of which include decreased appetite and weight loss.

More disturbing, Beauty was the forty-sixth peacock killed in the Lanes in less than two years.

The very next day, someone ran over a peahen on Buckskin. Peahens don't lurch into traffic, and it's impossible to drive fast enough on Buckskin not to notice a large bird in the road. That killing seemed intentional, though perhaps aggressive accidents happen. But the one two weeks after that? When someone walked up on a peahen docile enough not to be spooked and fired a six-inch metal bolt from a mini-crossbow into her back?

That was definitely on purpose. There were peacock killers, or a single serial killer, loose in the Lanes.

Peafowl arrived in Palos Verdes the same way they arrived everywhere outside of India, which is that a rich person brought them. Frank A. Vanderlip, Sr., a banker from New York who, among other things, helped design the Federal Reserve, bought sixteen thousand undeveloped acres on the peninsula in 1913. He was a serious bird fancier, the sort who built enormous aviaries on the land below his cottage and employed a full-time bird doctor. Vanderlip, in turn, most likely got his peacocks, two males and four hens, as a gift from one of the daughters of a man named Lucky Baldwin.

Lucky Baldwin introduced peacocks to California in 1879, six India blues imported from the subcontinent and set loose on Rancho Santa Anita, eight thousand acres he bought four years earlier for two hundred thousand dollars, the first payment for which he made in cash counted out from a tin box. He was an ideal importer: Only a person such as Lucky—not simply wealthy but flamboyantly so, a character invigorated by the glaring limelight—would have indulged in such birds.

His real name was Elias J. Baldwin and he hated being called Lucky (which pretty much everyone did) because it sounded like he hadn't earned his fortune with his wiles and his grit. Lucky considered himself a self-made man, and that was mostly true; he was an Ohio farm boy who made a profit on the wagon train to California by selling brandy, tobacco, and tea to the Mormons in Salt Lake City. He made millions investing in gold and silver mines, and he owned luxury hotels and Thoroughbreds and almost fifty thousand acres of land, much of it among the most fertile soil in the San Gabriel Valley. He also liked to gamble and carouse and chase women, and none of that was a secret: He'd been married four times, twice to brides who were only sixteen, taken to court more than once by jilted lovers, and shot in the arm in his own hotel by a woman who said she was his cousin and that he "ruined me in body and mind." Playing the villainous cad was part of his brand, and he dressed the part, too, with a black hat and black coat and bushy silver mustache he wore like a costume.

Indeed, that was basically his defense when a young hairdresser from Pasadena sued Lucky in 1896 for what was called, scandalously, seduction. Lillian Ashley, who was originally from Boston and forty years younger than Lucky, claimed that the old man, one night in his hotel, jotted a wedding contract on a piece of stationery and swore it was as binding as a church wedding. They both signed it and spent the night in her room in his hotel. A week or so later, she discovered her purported husband was already married, and not quite nine months after that, Lillian gave birth to a daughter. She named the baby Beatrice, and then she sued the old man for seventy-five thousand dollars, which, to add some perspective, is about $2.3 million in today's dollars.

Lillian's basic claim, that they'd had sex in his hotel, was not in serious dispute. Lucky just didn't want to pay her off. He wasn't in the business of giving away money. He would stiff the lawyer defending him and an expert witness, too. Quite the archetype, Lucky. But his defense was viciously simple: Lillian Ashley was, in the euphemism of the day, an adventurous woman. "My public reputation," he was reported to have once said, "is such that every woman who comes near me must have been warned in advance."

Near the end of the trial, when Lillian was on the stand, her little sister Emma, who'd been quietly reading her Bible on a bench at the back of the courtroom, slipped toward the front and took a seat at the gallery rail directly behind Lucky. No one thought anything of it. That's where Emma had sat through all of the testimony, never making a sound, just reading her Bible. She was in the Gospel of John that morning, thirteenth chapter, verses twenty-six and twenty-seven, the moment when Satan enters the heart of Judas Iscariot.

Emma pulled a revolver out of her handbag and held the barrel not two inches from the back of Lucky's head.

Then said Jesus unto him, That thou doest, do quickly.

She pulled the trigger, but she jerked it, tipped the gun up, put the bullet in a wall, and left a powder burn in what was left of Lucky's white hair. A bailiff hauled her out as she sang "Nearer, My God, to Thee."

"It would not have been murder," she said from the jail. "It would have been retribution. I tried to kill that man because I believed it my duty before God to rid the world of the wretch who seduced my innocent sister and dragged her down to the lowest depths. I believed it to be God's will that he should die by my hand, but it was not to be and I accept His will."

A jury took only two minutes to decide that Emma was temporarily insane when she pulled the trigger, a verdict that appeared to be concerned less with her actions than with her target. A judge, meanwhile, ruled that Lillian was not entitled to any compensation from Lucky because, as Lucky's lawyers had argued, she had indeed been an adventuress.

Lucky lived for another thirteen years. His holdings in the end were so vast that all or part of fourteen San Gabriel municipalities—Monrovia, Baldwin Hills, Sierra Madre, West Covina, and so on—would be built on them. Lucky incorporated one, the city of Arcadia, made himself the first mayor, and built a horse track, Santa Anita Park.

His original six peacocks multiplied year after year. There were at least fifty of them on the ranch, and they killed the snakes and ate the snails and raised a honking ruckus when predators skulked about. And those birds continued hatching more peacocks, and now, more than a century after his death, descendants of Lucky's birds still roam his former lands. Hundreds of them, too, the largest enclave at the Arboretum in Arcadia, which incorporated a peacock into the city logo. There are smaller musters in Pasadena and Glendora, and farther southwest, in places Lucky never owned, like the Palos Verdes Peninsula. At this point, it's impossible to determine which birds are directly related to the Rancho Santa Anita flock, but that's not terribly important. Lucky introduced peacocks to Southern California, and it is not unfair to consider all of them part of his legacy, and possibly the most visible and enduring part.

Which is fitting. They are a gift in one sense, so much beauty and magic. Yet they are also noisy and entitled and leave distasteful messes for others to clean up. Shrieking showboaters, if one is so inclined to consider them as such, taking over whole neighborhoods, shitting all over the place. Wretches, really, deserving of retribution, and certain people will think like that long enough and hard enough until one day they're Emma Ashley, slipping up from behind and hoping no one notices.

During the Depression, Frank Vanderlip, Jr., gave all of his birds to William Wrigley, Jr., who made a fortune in chewing gum and had a bird park on Catalina Island. Except the peacocks. Those Vanderlip let roam on the peninsula. Over time those birds made more birds, and the flock spread beyond the estate, which for many years was mostly undeveloped chaparral and scrub canyons: Palos Verdes Estates, incorporated in 1939, was the only municipality on the peninsula until Rolling Hills and Rolling Hills Estates came along in 1957 and, in 1973, Rancho Palos Verdes. By then, people couldn't help but encroach on established peafowl territory. For many, it was part of the charm; indeed, in the early sixties, the mayor of Palos Verdes Estates was suspected of loosing his own peacocks on the peninsula.

Not everyone was charmed, however. Peacocks are noisy in the breeding season, and several dozen of them in close proximity are extremely noisy. Peacocks and peahens also poop prodigiously, and they enjoy young, tender plants and dust baths that may or may not be in a groomed planting bed. The males pick fights with their own reflections in the sides of well-polished cars, and the feet of either sex will leave scratches in the paint. Palos Verdes Estates had a management plan to keep the flocks in check with humane trapping and relocation as early as 1986, and the other cities over the years restricted the birds to certain neighborhoods and periodically hired trappers to remove a few dozen.

For decades, there has been a low tension between those who adore the peafowl and those who find them a destructive nuisance. That tension is almost perfectly balanced between the pro-peacock people and the anti-peacock people: Neighborhood surveys about whether the flocks should be thinned or left alone have almost always come back evenly split. So there was a kind of détente, not quite a peace treaty but a cessation of open hostilities.

Until the birds started dying.

The notes concerning the first dead bird on the list are sparse because at the time, which was May 30, 2012, it seemed to be of no particular significance. The death was recorded only as a bird, sex not specified, found dead in an unknown location by an unknown person, who called animal control to pick up the carcass. It happens. The peacocks and peahens in the Lanes are feral. Animals die. A dead squirrel doesn't make much of an impression, either.

But a peacock turned up dead the next day under a bush in Virginia Gerisch's backyard on Buckskin Lane. Two days later, Virginia, who goes by Gini, saw another one in the yard across the street. Her neighbor said it just fell out of a tree. By the end of July, ten more peacocks were dead on Buckskin Lane, and four of them were on or near Gini's property. Neighbors, ones who like the peacocks, packed six of the bodies in ice and took them to a veterinarian for a postmortem. All six had been shot by pellets or BBs.

At that point, it seemed prudent to mark when the killing spree began.

That was also about the time, midsummer 2012, that Captain Cesar Perea got involved. He's a cop (a lieutenant then) with the Society for the Prevention of Cruelty to Animals Los Angeles. A real cop, too, a former sheriff's deputy and police officer in San Diego who wears a badge and carries a gun and has the authority to arrest you for felonies, which people sometimes don't understand because animals typically are not associated with serious crime. In fairness, much of Cesar's job involves education, like, say, reminding people that it is illegal in California to keep your dog tied to a fixed object for longer than three hours. But he also investigates nightmare incidents, such as a dead possum, beaten and stabbed and burned and left hanging from a noose on a chain-link fence, the seriousness of which is less about justice for a dead marsupial than it is about curtailing the person who tortured it to death. People who inflict such cruelty on animals are often just practicing. The connection between animal abuse and domestic violence, child abuse, violent crime, and full-on sociopathy is well established; Ted Bundy, Jeffrey

Dahmer, and Albert DeSalvo all maimed and killed animals before they started on people. And Cesar's cases are investigated with the same vigor and forensic sophistication as any with a human victim: In the possum case, he used DNA evidence lifted from the rope to find a young felon and his unsettling collection of homemade daggers, swords, and hatchets.

"We jokingly say we're the pre-crime unit," Cesar told me, "because we're gonna get you before you start doing this to people."

The peafowl killings slowed after Cesar showed up. An appearance by a law enforcement officer can have that effect, especially if he takes off his jacket at a community meeting and everyone can see the gun he carries, which isn't meant to intimidate but does remind those in attendance that murdering peacocks is a serious felony. Just two birds were killed in August, a peahen on Dapplegray and a peacock in Gini's yard, both shot with pellets, and then only one in each of the last four months of 2012. January and February passed before the next one turned up, shot with a BB and crawling with maggots at the head of the trail leading down to the riding ring. Gini's house is next to the trail.

Dead peafowl came in clusters after that. Eight in the summer and fall of 2013, six of which were run over by cars. Three were shot with BBs in October, including one that died in Gini's yard. In November, five were poisoned, though by what was uncertain; four of the bodies were found on Gini's property and the fifth was on the trail. "I was the pet cemetery, apparently," Gini said. She's lived in Palos Verdes since 1972, and in that particular house above the riding ring for decades. She used to keep a flock of chickens in the yard, and the peacocks learned they could poach their feed; there'd be two dozen of them outside Gini's house sometimes, taking dust baths in the yard and roosting in the pepper tree. A couple of people in the neighborhood set out food just for the peacocks, and the people who don't like peacocks like the feeders even less.

Cesar focused much of his investigation around Gini's property. He installed surveillance cameras and lingered on horseback and concealed himself in the greenery at the edge of the trail. "I thought for sure we'd be laying in the bushes doing surveillance and I'd see someone come out and do the act," he said. But he never did.

The killings picked up again in the spring of 2014, when Beauty and ten other birds died. Cesar thought he caught a break not long after, in July, when a middle-aged guy in a silver Mercedes stopped on the side of Eastvale Road, pulled an air rifle from the backseat, and shot a peacock in the throat. A teenager saw the whole thing. "He basically did a drive-by on these birds," Cesar said. "That couldn't be the first time he'd done it. I mean, who drives around with an air rifle in the back seat of a hundred-thousand-dollar car?"

The kid got a good enough look for a police artist sketch of a square-headed generic white guy in aviator shades. Cesar ran down all the silver Mercedes sedans, even the loaners out from the dealers, knocked on all the doors. All dead ends.

And then the killing was over. "It just completely stopped," he said. "Went away for the longest time." Maybe. Cesar visits the Lanes every now and again. He's never done a formal count, but in 2018 he was sure there were fewer peafowl than there had been four years earlier. Kathy Gliksman and her daughter Mary, who kept records on all the killings, thought so, too. "So the other thing," Cesar said, "is that maybe whoever was doing it just got better and, you know, went underground."

. . .

The people who are opposed to the peafowl in the Lanes, or at least to large numbers of them, do not have irrational complaints. They aren't imagining the noise or the damage or the droppings. They are perhaps more sensitive to those things, but that does not make the squawks in the night any less loud. Still, theirs is not a sympathetic position, for two reasons.

The most obvious is that the peacocks were there first. Anyone who has bought a house in the Lanes during the last sixty years knows that. The birds aren't hiding, waiting to spring out screeching only after the deed is signed. They're right there in the open, standing in the street, perching on fence rails and rooftops, picking bugs off the lawns. That's the appeal: Between the peacocks and the horses, the place is like an oversugared storybook. "People think it's just the cutest thing ever," Gini said. "Until they're here for a day. It's certainly something you have to get used to, but I equate it to moving next to a busy street or the train tracks. To me, what you do then is you move." Moreover, the fact that peacocks are noisy and poopy isn't a secret. Really, it's a bit like moving to the beach and complaining about the sand.

The other reason the opponents get little sympathy is that they are criticizing peacocks. If wild turkeys were tearing up gardens and kicking off roof tiles, odds are no one would much mind culling the flock. An infestation of marauding raccoons would be trapped at will, and likely with enthusiasm. Rats? The grocery store sells devices and toxins to kill them in all sorts of painful and barbaric ways. If one were to speak out against, say, flocks of beige-and-black Canada geese clogging Buckskin and Sorrel, spectators would nod and murmur and concede that those were all reasonable points.

But peacocks? The ones colored an unreal blue, feathers like the fronds of an especially exotic tropical plant? Most people are charmed by peacocks, or at minimum are pleasantly neutral toward them. Peacocks are beautiful, and pretty birds can get away with anything.

In fact, one of the locals who used to be quoted quite often saying not unreasonable things about how reducing the number of peafowl might not be a bad compromise declined to talk to me. "Do you know how much shit rains down on you when you get pegged as anti-peacock?" that person said. "You might as well hate puppies and babies while you're at it."

For all that, it is entirely possible to actively dislike peacocks and also be appalled when they are tortured to death. The birds in the Lanes were not humanely killed (which, for the record, would be a felony). Most of them were injured and left to die lingering, excruciating deaths from lead poisoning or festering wounds or shock and exhaustion. The ones who were killed quickly were also killed painfully, deliberately crushed by tires. Two birds, a peacock who survived and the peahen who died, were shot at close range with bolts from a mini-crossbow. The cruelty was not subtle.

Cesar stayed with the case for more than two years, collecting DNA and pellets dug out of dead birds and running license plates and hiding in the bushes, in part because of that cruelty. There was no practical reason to kill sixty birds. The numbers would not appreciably decline one pellet-shot peacock at a time. The noise would not abate. The males would not whisper among themselves to be more careful about dinging certain cars, and the females would not be more selective with where they incubated their eggs. The killings were an exercise in sadism, nothing more.

Rage killings would at least be understandable. Some guy finds a peacock clawing his brand-new BMW and he swats it off the hood with a golf club that happens to crack the bird's skull, that sort of thing. Awful and hideous and indicative of poor anger-management skills, but understandable. Those typically would be one-offs, too. Most human murders are committed in the heat of a single deranged moment of anger or panic, and such moments almost never present themselves more than once in a lifetime.

But these were not that. These were vengeance killings, an altogether different category. Colder, meaner, more controlled. Vengeance is what rage becomes when it's compressed over time, like the way carbon transforms into a hard diamond. Vengeance is cultivated, nurtured, embraced. Rage is a brief in the newspaper; vengeance is on the front page.

Killing peacocks in the Lanes required preparation, and preparation does not occur in a state of rage. Perhaps a few birds were run down in spasms of anger, and those deaths were cruelties of opportunity. But the others were preceded by multiple steps, each of which involved a deliberate decision to continue. There are logistics to contend with, buying the pellets and loading the gun and scattering the poison, and often some amount of labor. Neighbors in the Lanes remember the man who, before the big rash of killings, trimmed branches from the tall tree behind his house so that when the peacocks roosted on the few that remained, he had a clear shot at their silhouettes against the western sky. "Looked like something you'd see at the cowboy shooting range at Disneyland," one of the neighbors said. For whoever was killing birds repeatedly, there was, after a time, the additional complication of avoiding Cesar Perea and the surveillance cameras and dodging anybody who wasn't winking and, therefore, complicit.

Finally, there was figuring out how to get all those carcasses on or near Gini Gerisch's property. It could have been a coincidence, all the birds that used to scavenge her chickens' feed turning up dead for her to find. The pellet through her garage window could have been an accident, too. Probably not, though.

"My hunch," Cesar said, "is this might have started out as a kind of 'screw you' to a neighbor. But it gave him a little thrill. So now he's still got the 'screw you, neighbor' part, and he's sticking it to the birds that damaged his property or pissed him off, and he gets another thrill."

Cesar doesn't know who *he* is, or if there's only one. He might never know. The point is, a person who gets a thrill from scaring his neighbors and torturing animals is operating far outside the range of acceptable human behavior. And over peacocks, which seems almost absurd. How could a creature universally admired for its appearance create such rage, generate a thirst for vengeance? Whoever is killing them is risking prison, and for what? The birds can't be exterminated by a slow, surreptitious slaughter. So why?

Gini figured it out. "People think they're cute," she told me. People drive into the Lanes and it is fairy-tale land. They see in those birds what I saw, elegant hallucinations on a fence rail, cobalt sylphs rising from the dust. They offer, just by standing there, a swirl of wonder, a glimpse of fantasy.

And then they go and act like birds, whooping and pooping and trashing the garden. To a certain kind of person, it feels like a bait and switch, as if they've been betrayed. It's the stuff of pulp fiction and tabloid crime, beauty and betrayal, and it always ends badly.

Chapter Fourteen

Emmett had a day off from school and wanted to visit Carl, who was well into his second week as a patient at the Birdie Boutique because he still had pebbles in his belly and metallic spots in his X-rays. Our allegedly mythical bird had proved himself to be maddeningly mortal.

Emmett hadn't seen Carl since Uncle John and I threw a net on him. He brought some quartered grapes with us to Burkett's office, and he poked one through the bars of the big crate where Carl was sequestered. Carl looked at it but did not move. Emmett dropped it, trying to land it on the grate, but it fell through. So did the next three.

Two cat carriers were stacked next to Carl's cage. The bottom one had a barred rooster that crowed every six minutes or so, and the top one held a chicken with a badly infected foot swollen up like a catcher's mitt. When I first met Burkett, he told me he would make a whole practice out of chickens if he could; they're smart and social but stoic, hard to diagnose sometimes, a challenge for a vet. And tough. He'd sutured roosters without anesthesia.

The one next to Carl was very loud.

On the other side was a green parrot, which one would know only from his head because the rest of him was wrapped in white bandages. Directly above was a blue-and-gold macaw who had plucked all of his feathers from the throat down. His body was completely naked, as if someone had put a macaw puppet head on an unwrapped Cornish hen. There was also a cockatoo with a crooked beak and, in the cages across from Carl, a yellow cockatiel that rapidly paced the floor of the cage like a stir-crazed inmate, and a chatty African gray parrot that alternated between "Come here" and "Oh shit."

"Kinda cramped, isn't it, pup?"

Emmett nodded but didn't say anything. He was focused on getting Carl to eat a grape.

"What do you think we should do, all things considered?" I wasn't soliciting advice, just trying to gauge his mood.

"Save Carl," he said, working another grape through the bars.

"Don't worry, we're gonna save Carl."

"We have to. He's our pet. He's our responsibility." He looked around at the other birds, Carl's neighbors, sizing them up. "And he's a peacock. Everybody loves a peacock."

I muted a laugh. "Yeah, well, you'd be surprised."

"Who doesn't like a peacock?" Emmett suddenly wanted to know.

I decided not to tell him about Palos Verdes or any other gruesome episodes I'd come across. A few weeks earlier, I'd gone to Hawaii for a story about a false alarm involving an incoming ballistic missile. No one died, but more than a million people believed they would be dead in forty minutes or less, and I was interested in how people would spend those last moments. While I was there, I met a lawyer named Earle Partington who once defended an old woman after she beat a peacock to death with a baseball bat. There are peacocks all over the Hawaiian Islands. A New Zealander named Francis Sinclair brought the first ones in the 1860s—Princess Ka'iulani adored them; there is a statue in Waikiki of her feeding one—and there are some now in zoos and botanical parks. But there are also feral colonies, including one in a complex of condominium towers in Mākaha, on the northwest coast of Oahu. Most of the residents tolerate them, and many even enjoy having them.

Sandra Maloney, however, damn near lost her mind. She said she was sleep-deprived and depressed from the constant squawking and hooting and pooping. One afternoon in May 2009, near the barbecue grills, she grabbed a male by his train and swung at him with her bat. She missed, swung again, and connected with his head. The kill was not clean: The peacock stumbled around, one eye dangling out of the socket, and tumbled down a few steps before it died. Maloney said she was going to cook it, but a security guard called police before she could get the bird up to her condo.

She was charged with misdemeanor animal cruelty, pleaded not guilty, and hired Partington to defend her at trial. He argued to the jury that peacocks are pests and that there is no Hawaiian law against killing pests. "They're essentially vermin," he told me. "They're rats with feathers." (That is not an uncommon position: Thirteen months after Maloney bludgeoned one to death, Honolulu officials had eighteen of the birds killed at the Koko Crater Botanical Garden. "The purpose of the botanical gardens is to preserve native plants," a spokesman for the city said, "not provide a home for nonindigenous peafowl that are noisy and relieve themselves everywhere.") The jury found Maloney not guilty in less than two hours.

The law at the time was vague enough that, as a legal matter, it was okay to beat a peacock to death in Hawaii. Seeing as how most people find such behavior abhorrent, the law has since been tweaked to make clear that animal cruelty statutes apply to all living creatures, even pests and vermin.

I had plenty of other material, though. "That's an interesting question," I said to Emmett, having remembered something not bloody. "I was reading about this place in India where the farmers talk about 'the peacock menace.' It's their national bird and it's sacred and everything, but they're wrecking the crops."

Emmett looked skeptical. "Where?"

I searched notes on my phone. "This place, for one," I said, pointing at Punjaipuliampatti on a map. Wild peacocks can strip forty percent of a field's crop yield, I told him. "Trust me, it's true. So not *everyone* likes a peacock."

The African gray said "Oh shit" in a muted bird voice, and then the rooster crowed next to my ear like a siren.

"Carl isn't a menace," Emmett said when the racket subsided. "He's *Carl*." About that, he was spot-on. To be a menace, Carl would need to have much more going for him, starting with a full set of proper feathers and higher self-esteem. Carl in that little cage with his one keeling ocellus could just about break your heart.

A week later, Mr. Pickle inched toward me, stutter-stepping, moving one foot forward and back like a short loop of video. This was a new development. I had a blueberry at the tip of my fingers, and I was watching him from the corner of my eye. Maybe if I didn't look at him, I thought, he'd see me as less of a threat, though really, that shouldn't have been a concern anymore, even for a bird.

The berry rolled off my fingers. Mr. Pickle jabbed his head down, grabbed it, took two quick steps back.

Why Peacocks?

I reloaded, held my arm out again, turned my head. Ethel was watching, shifting her stare from my face to my fingers and back. I'd always thought she'd be the brave one to take the first treat from my hand.

Mr. Pickle twitched in my peripheral vision.

I felt a firm peck on my fingertip, and the berry was gone.

Not a triumphant moment, exactly, but a satisfying one. Anything one tries to accomplish for seven months and twenty-four days will, in the end, at least be that: satisfying. I was going to share this development with Dr. Burkett later that afternoon, when I dropped off more blueberries and food, but I suspected he might not share my satisfaction. The more I thought about it, the more ludicrous it seemed, like I was a man with far too much time to kill. Martha Stewart wouldn't waste that many hours coaxing a bird to take a berry from her hand. To have peacocks was probably enough; she didn't need them to bond with her.

I said hello to Julie at her desk, and she waved me toward the cage room. "Hey," Burkett called out cheerily when he saw me passing the doorway of the OR. He was seated at the far end of the operating table and was wearing a dark bandana printed with feathers tied around his head, and glasses with magnifying lenses and a bright light shining from the bridge, like any surgeon. "We got your bird in here."

That was morbidly obvious. Carl was on his back on the operating table, legs up, wings flopping out to either side like long, warped shelves. His head hung down from the table, and a tube connected to a tank of anesthetic gas was taped to his beak. Two nurses monitored his vitals and retrieved tools and otherwise assisted. Blue surgical paper covered most of Carl's breast and neck except for an opening at the base of his throat, where Burkett was starting an incision. I worked my way around to his end of the table, tentatively at first, like Mr. Pickle stalking a blueberry, until I was hovering over Burkett's shoulder. I'd never seen a bird cut open that wasn't already dead.

Burkett made a hole about the size of my thumb, through which he threaded a rubber tube about eighteen inches long. Then he used a syringe the size of a turkey baster to squirt water through the tube and into Carl's gut. "Lavage," Burkett said. "Washing out his stomach, see if we can force some of those stones out."

Water sputtered back out almost immediately, as clear as it was going in. Burkett forced more through the tube, caught the return gush in one of those stainless-steel cat dishes. On the fourth or fifth squirt, a few tiny pebbles and undigested food trickled from the tube. By the seventh, the liquid was running a flat forest green, mostly acid and bile, I gathered.

Burkett was trying to hold a syringe, a tube, and a steel bowl all at once. He bobbled the bowl, tipped it a bit, and I snapped a hand out, steadied it to catch my poisoned peacock's stomach juices. It crossed my mind briefly that Burkett and two professional nurses probably did not need my help or, for that matter, want an untrained hand suddenly darting into the surgical arena. But I chased that thought away. I'd trained chickens to jump, by God, and a peacock to eat from my hand, and now I was assisting in an avian procedure so complicated it had a French name. *Lavage.* I'd need to remember that word when I told the story to the boys.

The lavage water was coming back up mostly clear again. The big rocks were still down there. For the next four hours, Burkett rooted around in Carl's stomach with a long cable that had minia-ture grabbers and a teeny camera at the end so he could see what he was doing on a computer screen. The blue surgical paper had fallen away, and the opening in Carl's throat had expanded, a wet red gash against the deep blue-green of his breast. When Burkett inserted the grabbers, the light from the camera made Carl's insides glow red, like a feathered jack-o'-lantern that had been carved with very little imagination. It was all very high-tech and labor-intensive for a bird who had poisoned himself in a garbage coop.

No one told me to move, so I didn't. I texted Louise. *Operating on Carl. Very cool. I'm helping!*

Burkett pulled out as many rocks as he could, which was most of them, and then decided to sew up Carl's throat. I assumed some of the feathers on the periphery of the incision would get in the way, so I silently volunteered my hand to hold them at bay. No one objected. As Burkett was putting in the final stitches, I asked how long it would be before Carl was up and around. "As soon as the anesthesia wears off," he said. He was sewing, so he didn't see me raise a dubious eyebrow. "He's a bird. Birds don't get to recuperate. In the wild, they'd get eaten."

Carl started waking up almost as soon as the gas was turned off. There were a few little stones left in his stomach, but there were no bright points of metal when Burkett X-rayed him two days later, and his blood work came back clean. On the first Sunday in March—three weeks after I brought Carl in and twelve days after Emmett tried to feed him quartered grapes—Burkett called me to come get my bird.

Calvin went with me, a mix of Sunday-morning boredom and a twelve-year-old's curiosity. The door was unlocked when we got there, but lights in the front were off because technically the place was closed. We found Burkett in the operating room hunched over a caique, a bird approximately the size of Carl's head, which made me realize he must have an exceptionally precise and delicate touch.

There was a different nurse working with him, a sandy-haired woman about my age. Her name was Valerie.

"I heard about your bird," she said, glancing away from a monitor just long enough to make eye contact. "The poor baby. You need some sand."

"Sand? For what?"

"For his pen. Put down four or five inches of sand, and he'll have something nice and soft to walk on, and he won't be able to scratch anything else out of the dirt. And all the liquid, it filters down through the sand so you don't have to deal with it."

"Oh my. That's brilliant. How did I not think of that? It'd be like a giant cat box."

Valerie smiled and nodded. "And the liquid," she said again, "it all . . . filters . . . down." Another nod. "And sprinkle a lot of salt on the ground first. That'll kill any parasites that might try crawling up."

Calvin nudged me. He wanted to get Carl and go home, not talk peacocks with a stranger. We thanked both Valerie and Burkett and went looking for Carl in the room with avian patients in cages and crates. The naked macaw was still there, but Calvin's eye caught Carl's bright blue breast first. He smiled, slid down on one knee, said, "Hey, buddy, it's us." Carl did not move, but he blinked.

It was at this point that I realized I had no idea how to get a convalescing peacock out of a cage and into a feed bag. *If he catches you good,* Burkett echoed in my head, *you're gonna bleed* a lot.

I sheepishly took a step back, leaned in to the OR. "Um, Doc? How do I do this again?"

His head bobbed with a small laugh. "Valerie?"

"Of course, darlin', I'll help you." There was a rasp in her voice and a maternal warmth in her accent.

Why Peacocks?

The trick, as I witnessed, is in speed and authority. Valerie bent down, opened Carl's cage, and, without hesitating long enough to blink, reached one arm across his back, pulled him in tight to her side, grabbed his feet with her other hand, and stood back up. Carl made one instinctive thrash. "Settle down," she told him in a tone that was soothing and firm all at once. Carl obeyed.

"That was pretty badass," I muttered to Calvin.

"You learn. I've got five peacocks, a boy and four girls. Always happy to help. So call me if you need anything."

Peacocks. *Boy* and *girl* peacocks. I liked her.

I looped twine around Carl's ankles, just below the spurs, the way she told me to, and then held a Mule City bag while she maneuvered him in. His crest poked from the hole I'd cut, which I immediately realized was the same dumb mistake I'd made once. A properly bagged bird is subdued and defeated, but Carl was breaking for daylight. His head was out, followed by his entire neck. The hole tore slightly but held at his shoulders.

Valerie wrapped a length of nylon rope around the sack, pinning Carl's wings and feet to his torso. "Now, one of you is gonna have to hold him," she said, looking at Calvin, "and I'm guessing that's you." He laughed nervously. "No, really," she said. "You have to hold him."

In the parking lot, Calvin situated himself in the passenger seat while I held our trussed bird under my arm, not squeezing, exactly, but with a notable stiffness. "Ready?" Calvin nodded apprehensively. I slowly lowered Carl, holding him now with both hands, onto Calvin's lap. "Put your hands right behind mine," I said, "and push down, but not too hard. Just hold on to him." He moved mechanically, rigidly, and the knuckles in his thumbs whitened.

I closed the door gently, as if a rough bump might startle Carl into a slashing panic. Walking around to the other side of the car,

I wondered if this was terrible parenting, endangering my child by forcing him to hold a large wild animal. Then again, Carl was not exactly a honey badger. Actually, he was sitting as I'd initially imagined a bagged peacock would, his body restrained in some mysteriously efficient way yet his expression calm and regal, refusing to concede he'd been bundled up like produce.

Inside the bag, Carl was shitting profusely. So much that it leaked through the Mule City weave and onto Calvin's leg. But the boy did not move. He kept his eyes on the back of Carl's head, as if he was watching for the early warnings of a bird frenzy and wasn't going to be caught unaware. I quietly declared this to be an enriching experience, a boy wrangling a peacock with his father. He relaxed his grip enough to give Carl a cautiously comforting pat with two fingers, and a burp of pride bubbled in me. My son was focused, calm, exactly the kind of partner a man needs when he's transporting a shitting, sharp-taloned peacock in a hatchback.

Chapter Fifteen

The sixth Baron Moncreiff stalked the pebbled drive in front of his castle, waving a nubbin of fruitcake and calling his peacocks. He was hollering "C'mon," but between his accent and his cadence, it came out as a single rounded syllable booming across Scottish fields that bent away into the distance. He bellowed from in front of his castle for a minute or so, shambled across the lawn toward the hedges to try over there, then returned to the drive.

He did not look like I thought a baron might. Nor did he look like a baronet—higher than most knights, lower than a baron—of which he is the sixteenth in a line winding back to 1626. Rather, Lord (as barons are somewhat confusingly addressed) Moncreiff— or, less formally, Rhoderick—had the rumpled insouciance common to people who've been landed gentry for so many centuries that they no longer fret about keeping up appearances. He was wearing, on this bright midday, corduroys that had gone shiny at the knees and a shapeless faded green sweater, and his white hair spiraled from his head in a shaggy approximation of a monk's ring. He seemed just a kindly man of certain years who enjoys, among other things, feeding his birds.

That was why I had come to see him at his castle, which did not look any more like a castle than Rhoderick looked like a lord, even though it was right there in the name, Tullibole Castle. It is more of a moderately large and very old stone house decorated with the requisite accoutrements of a castle: a small tower, turrets pasted to the exterior walls, and what looks like a machicolation, which is an opening high up through which boiling oil could be poured upon marauders. Inside, a narrow stairway winds up to the great room, a boxy cavern the size of a community theater and bone-cold because the fireplace is unlit, and there are portraits on the walls of people who appear to have lived in several different centuries. "Ancestors and such," Rhoderick told me, though with no enthusiasm to elaborate; the house had been in the Moncreiff family more than three hundred years, and it would have taken the better part of a week to get through all those biographies.

It occurred to me that Tullibole was a castle in the same way that my home is a farm: a reasonable, manageable facsimile. The lands surrounding Rhoderick's castle are exponentially vaster, with fields and woods and a moat-ish stream; and the outbuildings are substantially larger and more interesting than our sagging barn and old brick smokehouse in desperate need of repointing. But I convinced myself the basic comparison, one of scale, was workable. We weren't so different, the baron and I.

Rhoderick suspects there were peacocks on the property long ago. "In a Jane Austen–y kind of period," he said. But the basis for that suspicion is a single drawing that's framed and hanging in the great room. The castle is in the center, flanked by towering oaks in full leaf. A woman and child, in what reasonably could be interpreted as Regency-era clothing, are walking toward the front door. A peacock and a peahen are in the left foreground, and another peacock is off by himself on the right. Rhoderick did not know, however, if that was an accurate representation of Tullibole or, rather, a landscape architect's interpretation of what *could* be at Tullibole. Apparently, none of those ancestors and such bothered to conduct a census.

But peacocks for certain had been on the grounds of Tullibole for at least fifty years, since Rhoderick's father, the fifth Lord Moncreiff, brought three to the castle in the nineteen sixties. At first he kept them in pens, long and narrow and peaked. "Imagine a twelve-foot Toblerone," Rhoderick told me. He eventually released those three and they became a dozen and then two dozen, the population dipping with predation—herons, disappointingly, are a menace to peachicks—and rebounding with successful hatching seasons. Rhoderick isn't sure exactly how many peacocks and peahens are loose on the grounds, there are so many.

Surely Rhoderick and his wife, Lady Alison, would have answers. Surely they would know what one does with peacocks.

"C'mon c'mon c'mon," he was still calling from the drive. His peacocks did not come with the vigor and glee of Comet and Snowball. But then I saw a glint of blue moving in the field beyond the gate, slowly walking toward us. Already I was learning. Peacocks required more patience than chickens.

Tullibole Castle is near Crook of Devon, in Kinross, a twenty-minute drive north of Dunfermline, which is an old city of dampness and stone across the water from Edinburgh. Scotland used to bury its kings and nobles in Dunfermline. Malcolm Canmore, the great king who founded a Scottish dynasty almost a thousand years ago, is buried in the abbey at the edge of the city center, as is his queen, the former Margaret of Wessex and the current Saint Margaret, who founded the priory that became the abbey and who was canonized in 1250. Robert the Bruce, who led the first war of Scottish independence in the fourteenth century, is there, too, except for his heart, which was taken on a crusade before it was buried in a different abbey. Five other kings are interred at the abbey, as well as the mother of William Wallace, a blue-faced version of whom Mel Gibson played in *Braveheart*. For almost four hundred years, Dunfermline was an important city in Scotland, the center of royal and religious authority, until the capital moved across the Firth of Forth to Edinburgh in 1437 and the Reformation swept through the following century.

The town never reclaimed that level of status or glory. It kept the remains of the dead and ruins of the palace next to the abbey and the nave of the twelfth-century church, but Dunfermline wasn't resurrected for more than two hundred years, and then it was as a workingman's town. Coal and textiles, mostly, and damask linen most famously, much of it woven on handlooms by men in their own cramped cottages. One of those was a man named William Carnegie, who worked in a dirt-floored studio below the single room he shared with his wife, Margaret. In 1835, she gave birth to a son, Andrew; eight years later, after they'd moved to a larger yet still modest home, she had a second son, Thomas.

The Carnegies lived near the edge of a sprawling woodland estate called Pittencrieff. It was the only green space in the city, the place where Malcolm Canmore built a tower and the ruins remained, but Andrew and his brother could not play there. No one could. Pittencrieff had been private property for centuries, and the current owners, ripening aristocrats by the name of Hunt, forbade the townspeople to enter. The public grievance festered for generations, but to a boy, the Glen, as it came to be called, was a wonderland. Forbidden and unreachable but a wonderland nonetheless.

Why Peacocks?

"When I heard of paradise," Andrew wrote years later, "I translated the word into Pittencrieff Glen, believing it to be as near to paradise as anything I could think of. Happy were we if through an open lodge gate, or over the wall or under the iron grille over the burn, now and then we caught a glimpse inside."

The Carnegies left in 1848 for America, where Andrew, who was twelve, grew up to become one of the richest men on the planet, possibly in the history of the world, depending on who's doing the accounting: At his peak, Carnegie was worth north of $300 billion in 2020 money (Jeff Bezos, the wealthiest man alive that year, was a piker at $204 billion). But Carnegie believed that accumulated riches were only held in trust for the public, that wealthy people had a moral obligation to distribute money for the public good, and to do so while they were alive. "The man who dies thus rich," he wrote in 1889, "dies disgraced."

He'd been giving away money for decades by the time he sold Carnegie Steel in 1901 and retired, at the age of sixty-six, to be a full-time philanthropist. He funded theaters and a university and twenty-five hundred libraries in the English-speaking world and more than a dozen major trusts, institutes, and endowments. Carnegie built one of those libraries in his hometown, and people in Dunfermline will tell you with understandable pride that it was the *very first* Carnegie library, that the cornerstone was laid by his mother, Margaret, in 1881, and that there was a grand public holiday when it opened two years later. Dunfermline also has a Carnegie Hall and the Andrew Carnegie Birthplace Museum and the Carnegie Leisure Centre and the Carnegie Conference Centre and the Carnegie Dunfermline Trust, which allows Carnegie money to be spent on other things that do not necessarily have the Carnegie name on them.

I learned most of this, initially, from a tour guide. I needed this primer because I had not come to Dunfermline for the abbey or the ruins. I was not aware of the kings and queens who were buried there, nor that one of the most influential men in modern Western history had been born there and had pretty much funded the place in perpetuity. (Nor did I know that the front man for Big Country *and* the bass player for Nazareth were both from Dunfermline, which is the kind of trivia I usually sniff out first.) I did not come for the history, fascinating as it is.

I came for the peacocks.

A few months earlier, I was searching for stories about peacocks living among people that did not involve half of those people complaining about the noise and the poop. Or killing them. Those stories are very hard to find, but I dredged up a headline from the local Dunfermline paper: "More Peacocks join The Glen Family!"

I had never seen an exclamation point attached to a peacock-related headline. Not favorably, anyway. There were references to "Dunfermline's iconic birds" and "the town's famous peacocks" and Tullibole Castle, which was supplying peacocks and peahens to the city. There was mention of a new children's book. There was not a single sour note in the story; even the comments were pro-peacock.

I emailed the reporter, told her I wanted to learn more about Dunfermline's famous peacocks, and asked if she could put me in touch with the right people. She forwarded my note to Jim Stewart, the chairman of the Central Dunfermline Community Council. *Hi Jim,* she wrote. *You can see below the bizarre request but can you help this guy?* (That was not an uncommon characterization of a request for peacock information, by the way.)

Jim reached out the next day. He would tell me everything I

wanted to know about peacocks and about Dunfermline, or he would find the people who could.

The most significant gift Carnegie gave Dunfermline does not have his name on it. In 1902, he bought Pittencrieff, which officially is Pittencrieff Park but the locals call it simply The Glen. The following year, he gave it away. "No gift I have made or can ever make can possibly approach that of Pittencrieff Glen, Dunfermline," he wrote in his autobiography. "It is saturated with childish sentiment—all of the purest and sweetest." He put the land in a trust along with $2.5 million in bonds paying five percent, "all to be used," he told the trustees, "in attempts to bring into the monotonous lives of the toiling masses of Dunfermline more of sweetness and light[.]"

One of the sixteen trustees was a wealthy mill owner named Henry Beveridge, and he's the person who introduced peacocks, a quartet, to the park in February 1905. There is no record that he imported the birds for the occasion, and no reason to believe so: It's just as likely he had them on hand at his house, which was called Pitreavie Castle, as peacocks were in fashion among people who owned big houses they called castles. Peafowl bones have been found in Britain dating back to the Romans, and the birds were an established status symbol by the Middle Ages. Chivalric knights purportedly swore an annual vow on peacocks (as well as swans, herons, and other birds), an idea they most likely picked up from an epic French Romantic narrative written in 1312 called, accordingly, "The Vows of the Peacocks." That a wealthy Scotsman would have a surplus dotting his estate six hundred years later is not a stretch.

In any case, the arrival of peafowl in Pittencrieff apparently was not marked with any exuberance. The very helpful staff at the Carnegie Library found just one contemporaneous mention, a single sentence at the end of a brief newspaper update in 1905 noting only that the promised birds had been delivered and would be confined until accustomed to their new surroundings. (This gave me a glimmer of hope about letting my own birds out, though that required ignoring the fact that Pittencrieff Park is seventy-two times larger than our yard and has a lot more land to which a bird could get accustomed.)

As their numbers grew, the peacocks occasionally were mentioned in meetings of the trust for their uncivilized antics, destroying plants and such, and the flock was periodically thinned by giving away a few birds. But for decades, peacocks and peahens had the run of The Glen and the city. They would patrol the cobblestones on High Street or the graveyard by the abbey, rest in a doorway or on a wall, bright blue ornaments against mossy, mottled granite. They never really bothered anyone. Well, except a barman everyone called Kill. He liked big American cars, had a Cadillac, black or green depending on who's telling the story, and very shiny. A peacock saw his reflection and pecked at it, as peacocks do, dinged the paint, and made Kill so mad he kicked the bird in the backside. Didn't seem to hurt him, though.

The point is, peacocks were a feature, not a bug, in Dunfermline.

Public budgets started getting tight in the early aughts. The little zoo in Pittencrieff Park—where there were donkeys and llamas and marmosets and tanks of tropical fish and a cockatoo named Billy who would take pennies from your fingers and hide them in gaps between the bricks in his enclosure—was shut down in 2002. The peacocks weren't officially part of the zoo, but they roosted on the building and were used to at least minimal care and feeding from humans.

Damian Williams volunteered to look after them. He was in charge of the toilets in The Glen, keeping them clean and stocked, and he has a soft spot for animals that runs in the family: His father had been one of the handlers in the little zoo back in the seventies and eighties, and Damian would chop up fruit for the sick animals and bottle-feed the baby goats. There were fifteen peafowl when he took over, but the job wasn't much work, just making sure the birds had access to food and shelter and weren't sick or injured. He was squeamish about syringing antibiotics into the white peacock after the vet patched up a wound, but Damian did it anyway because being the keeper of the peafowl, even part-time, was important. The peacocks had been there his whole life. In the summer, Damian collected train feathers for the day when he volunteered at a festival for disabled kids. "Hey, there's the peacock man!" kids would say, and he would pass out feathers and be happy because the kids were happy. He knew there was magic in those feathers, even if he couldn't say why. "If I see a dead pigeon or a dead seagull, I don't feel nothing for it," he told me. "But if I see a dead peacock? Aye, I feel something."

Then the birds started dying off. It was nothing Damian did or didn't do. Some got old. One flew into a window and broke his neck. Another one was found on his back behind the pie shop, and the half-serious joke was that he got fat from pie scraps and fell off the roof. A dog chased the white peacock into traffic from its roost on the wall along the edge of Coal Road. A few just disappeared.

By 2012, there was only a single peacock left in Dunfermline. His name was Clive, and he became a celebrity, a mascot, even. Nothing focuses a spotlight more tightly than being the last of your kind. Clive the lonely bachelor peacock made the national press that year, page two of one of the tabloids. The BBC mentioned him as well, in a story about how The Glen got partial funding from the lottery for improvements that included, ironically, a new enclosure for Pittencrieff Park's iconic peacocks.

But for the first time in more than a hundred years, there weren't any to enclose. Only Clive remained, walking around free and completely alone.

He had it coming, really," Jim Stewart told me. He meant the peacock who got run over on Coal Road, the white one. "Aye, he did."

He was smiling when he said it, setting up a joke. He has a round face and white hair clipped close and appears to be in a fairly constant state of restrained Scottish bemusement. "That bird used to sit on the wall along the road at night," he said, "and the headlights would hit him, and he'd look like a ghost peacock. Scared the hell out of people."

Jim, the council chairman, was my host in Dunfermline, and an excellent one at that. Every city should have a Jim Stewart. Jim took me to a hockey game one night in Kirkcaldy, where we watched the Fife Flyers lose to the Cardiff Devils, eight to three, and where we noticed that the title sponsors of the Devils, the company that paid to paste its name across the front of every jersey for the next three years, was a chain of discount fashion stores called Peacocks. We drove up to St. Andrews one afternoon to walk the edge of the Old Course and see the *Chariots of Fire* beach and noodle around a kilt shop, and we stumbled into a pop-up fair where there was a small trailer covered in familiar blue-green feathers called Screaming Peacock Gourmet Burger Bar. It did not serve peacock; the name and the decor were meant to stand out, as a peacock does. The special was venison burgers.

Why Peacocks?

On the way to a fish tea in Anstruther—that's takeaway fish and chips at teatime—Jim insisted that those peacock encounters had been coincidences. He is one of the reasons there are peacocks in Pittencrieff Park at all, but he is not obsessive about the topic. His efforts for the previous six years, and the efforts of several others, were about preserving the character of his city, which just happened to involve peacocks. The point was not to randomly dress out the city with gilded birds but, rather, to reclaim and preserve a place where those birds by right of tradition belonged. "When you walked through the glen, you were sure to see two things, peacocks and squirrels," he said. "If there's no peacocks, there's a little bit of history gone. And if you can have squirrels, why not have peacocks?"

I stared at him blankly. Every place has squirrels. There is no correlation, biological or mythological, between squirrels and peacocks. *If squirrels, then peacocks* is not a formula anyone has ever applied anywhere. "That's kind of a leap from squirrels to peacocks, yeah?"

Jim smiled. He has a droll sense of humor. The tagline at the bottom of messages that he sends from his mobile says, *Sent from my Aye-phone.* "No, not really," he said. "And what I mean is, if you can have *wow*, why not have *wow*?"

When the birds had almost all died off, Jim was the one who nudged a few reporters to write about Clive, pitched the lonely-bachelor angle. It wasn't technically true, though. A young peacock by the name of Malcolm was skulking about, but he was either hiding or was off somewhere else. In any case, no one ever saw Malcolm and Clive together, and the underlying theme, that Pittencrieff's peacocks were on the edge of extinction, was still true: Two confirmed bachelors aren't any more conducive to repopulating a park than one.

Sympathy for Clive produced two hens, donated in 2014 from an estate with a surplus. Volunteers named one for Carnegie's wife, Louise, which was at least one thing I had in common with the richest man in history; they named the other Henrietta, an homage to Henry Beveridge. Henrietta took a runner almost immediately, but Malcolm reappeared. A year later, a woman up the road in Crieff gave the park a white hen and five chicks. Pittencrieff Park was back to eight peafowl.

All five chicks died three months after they arrived, for reasons no one figured out, and their mother was put down a little more than a year later when the vets couldn't heal a festering wound. Four months after that, in July 2017, Clive died. It should not have been unexpected. He had arthritis in his legs and he was twenty years old, which is the ballpark life span for a wild peacock. It is just easy to forget an animal is wild once he gets his name in the newspapers.

While all those peacocks were being donated and dying, two things happened. The first was that the promised lottery money came through and the monkey cage at the old animal farm was rebuilt into a large aviary that is now impeccably kept.

The other was that Caroline Copeland went on holiday to the United States for her birthday. In the gift shop of a museum in Boston, she happened upon *Make Way for Ducklings,* the children's book about a family of mallards who settle on an island in the lagoon of the Public Garden. It's a classic book and one of those enduring emblems of Boston, like the *Cheers* bar or Bobby Orr: There is a statue of Mrs. Mallard and her eight ducklings waddling across thirty-five feet of cobblestone in the garden. Louise and I have a picture of Calvin straddling Mrs. Mallard when he was three or four, one hand on her head and the other on her neck, where the bronze has worn smooth from the hands of countless children.

"So I had a flick through it and thought, wouldn't it be lovely if Dunfermline had a book for the peacocks?" Caroline said. "I mean, how lovely, right? They've made this big thing for the ducklings, and we've got these lovely peacocks, and we hadn't done anything with them. And I just thought, if children read that story when they're little, the peacocks would mean something to them."

She wrote *Peacocks in the Glen Again*, about a young peacock who travels from Tullibole Castle to Pittencrieff Park. The community council helped her get it published in 2016, with all the proceeds going toward upkeep of the real peacocks, and copies moved so swiftly that a second book, *Christmas in the Glen Again*, was published the following year.

By the autumn of 2018, when Jim was telling me jokes about ghost-white birds, there were ten peafowl in Pittencrieff, including a troublesome fellow named Bruce rescued from a village where he was making a nuisance of himself. Six birds, a hen and five chicks, were a gift from Lord and Lady Moncreiff of Tullibole Castle.

A curious male paced toward the edge of Lord Moncreiff's driveway, the low December sun sparkling off the right side of the bird's neck, throwing a long shadow peacock to the left. Other birds, peacocks and peahens, were wandering out of the trees and in from the fields, cautiously inquisitive, like survivors of an unfortunate nuclear event.

Jim brought me to Tullibole, too, along with Suzi Ross, Pittencrieff's peafowl warden. He wanted me to meet Lord and Lady Moncreiff and, besides, a peacock tour of Fife isn't really complete until you hit a couple of castles. We'd already done Scone Palace, where kings were crowned for centuries and where one peacock got so belligerent that he was rehomed to Pittencrieff.

"When they see the others run, they run," Rhoderick said. He swung his arm in an underhand loop, released a piece of fruitcake. "But it's a 'Who's gonna run first?' thing."

The closest bird broke for the middle of the lawn, snatched a bit of fruitcake. "They pick up from the way the first one eats how good the food is," Rhoderick went on, throwing the last of the fruitcake across the grass. Another peacock followed the first, and then birds raced from the fence line and the hedgerow and up the slope behind the oak, all of them skittering about for crumbs while Lord Moncreiff brushed off his hands in front of his manageable Scottish castle. "They like the sun," he said, and he seemed pleased to have brought them into the light.

Eleven peacocks and four hens clustered on Tullibole's lawn. I had never seen so many in the open before; in Palos Verdes, they were constrained by houses and cars and horses. Here they had room to move for long stretches, and I had time to watch closely. For millennia, the peacock has been accused of strutting, of employing a haughty walk to harmonize with his purportedly haughty train and, for good measure, his haughty crest. It is a pejorative word, *strut*, much as *peacocking*, the verb, is pejorative. I had never noticed my birds to strut, but that didn't count. They were in a garbage coop and, presumably, demoralized.

Yet here, in the open, it was plain to see that a peacock does not strut. It walks like any large ground bird. A peacock, to my eye, walks much like a chicken, the head bobbing forward in rhythm with the steps, the movements mildly exaggerated because a peacock's legs and neck are longer, but not fundamentally different. The motion at slow speeds is almost the opposite of a strut, more of a mindful high step, as if the bird is avoiding a squish of mud or especially sharp stones. When they run, it is with the same silly exuberance as Comet and Snowball, their budding trains stabilizing, but not eliminating, the rolling hint of a waddle. Lord Moncreiff's birds, scampering for a ration of fruitcake, were not arrogant. They were hungry, and they just happened to be beautiful while racing for crumbs.

Lord Moncreiff was inviting me into his castle, but I really wanted to stay on the lawn. The December sun was warm on my back, and two blue dots were approaching from down near the stream, moving slowly over a wide green lawn, curious, deliberate. I wanted to wait for them, and I could have done so for hours.

Part Three

A PEACOCK IN AUTUMN

Chapter Sixteen

Five days after Carl came home, Emmett had another day off from school, a teacher workday, which is generally a parent not-workday. Louise was at her office and Calvin was at school, so I let Emmett sleep while I did the morning routine. I walked Tater, and then he sniffed around the bird pens while I let the chickens out—having learned months ago that they did not enjoy playing chase, Tater settled for a polite nod of a greeting—and checked on the peacocks.

That morning happened to be the first they'd woken up on sand. I'd had it delivered the day before, eight tons, and spent hours moving it into the pen with the guys who brought it, one wheelbarrow after another. None of the birds feinted toward the door, no matter how long it was left open.

The pen looked good, but my shoulders were sore and I had a vague feeling of shame. The world is running out of sand, a phenomenon so outlandish that I'd been mildly obsessed with it for a couple of years. Sand is the most heavily mined commodity on earth, the most exploited natural resource after water, and the demand, mostly for construction, has long outpaced legitimate supply. Sand thieves were stripping beaches in Morocco, and a sand mafia—*mafia!*—was killing people in India. And here I was, hoarding eight tons for three birds so they wouldn't eat poisonous trinkets that I was too lazy to dig out of the ground. Construction-grade sand, too, the good stuff that could have been used to build a school or an orphanage or a school for orphans.

It's almost physical sometimes, the effort to chase away the guilt when it comes creeping. "You're a neurotic fucker," my shrink friend, the one with the breathing exercise, told me once. "And I don't mean that Woody Allen dickhead bullshit. I mean you worry about shit you can't control. Stop it." He was not an orthodox therapist, but he was right. Emotional energy, like water, finds the path of least resistance, and it is always more convenient to worry about what you can't control because it's never your fault when it all goes to shit. My effect on the global sand trade, I reminded myself, was negligible to the point of irrelevant: If my peacocks weren't walking on those specific tons of sand, they would be under the pavers of a new patio or stirred into the concrete of another big-box store that peddled sick snakes. From that perspective, my use of it was not illegitimate.

Early that afternoon, Emmett and I were at the kitchen table, the parts of a radio-controlled monster truck arranged in front of us. It had been a birthday gift the year before, and he'd decided to use his allowance money to make a few upgrades, starting with the suspension, specifically the oil-filled front shock absorbers. He'd bought tiny aluminum caps to replace the tiny plastic caps on the shocks and a tube of slightly heavier oil to refill them. The truck came apart easily enough, but we were stalled because one of the caps wouldn't unscrew. It was stuck too tightly for Emmett's fingers to twist it, and mine were too big to get a grip. We kept at it, passing it back and forth, before I finally told him I'd get some needle-nose pliers from the barn.

I went out the kitchen door and across the driveway. At the corner of the barn, where stairs run up to the old hayloft, I saw a blur of movement. Something brownish, bigger than a cat, a mangy thing that bolted between the bottom and the second steps and ran toward the bushes beyond Cosmo's grave.

Since it was the middle of the day, it took me a moment to recognize it as a fox.

A queasy panic rose quickly. I looked in the direction the fox had come from, the far corner of the barn where a honeysuckle vine chokes the garden gate. There was a pile of barred black-and-white feathers. It was one of the girls, Comet or Snowball, though which one I couldn't tell because her head had been torn off and dropped two feet away. But there was only one. I called both of their names. They always come when called. I eyed the Japanese maple by the steps where the girls would fly up if they got spooked; I'd plucked them out of that tree more times than I could count. It was empty. I said their names again, more of a stage whisper this time because I didn't want Emmett to hear.

Oh, hell, Emmett. What if he came out to find me? What if he saw the feathers, the pieces?

I ran back to the kitchen door, looking around for the other girl. I suspected the fox had her. "Emmett?" No answer. I went farther into the house. "Emmett?" Shit. He'd already slipped out.

"Yeah?" he yelled from his room.

I called up from the base of the staircase. "Hey, I'm, uh, having a hard time finding these things. Just wait here for me, okay?"

"Okay."

I did a quick check of the bird pens. The peacocks were agitated, pacing. Ethel started honking, somewhat tardily, I thought. I looked behind the woodpile, in the old cabinets screwed to the back wall, in a steamer pot on a shelf, all places Comet and Snowball had hidden before. Then I followed Ethel's stare toward the bushes. The girls hardly ever hung around that area, but that was where the fox had run.

The fox was still there, behind the shrubs; I could see his fur at the edge of the branches. And the feathers of the other lady. He heard me and darted into the underbrush of the neighbor's yard, leaving the corpse behind. He'd eaten more of this chicken. I never found her head.

I felt sick, in that anxious way where my chest gets tight and my arms get tingly and light, as if they might float. I should have known foxes might be lurking about. Deer were as rampant as flies. There were raccoons and possums and squirrels and rabbits and voles and hawks and vultures and bats and mice, all of which we could see every day if we cared to pay attention. *Of course* there would be foxes.

Oh, fuck. I *did* know there were foxes, knew for certain, because I'd seen them before. It had been years, six or seven, early spring. I was in the garden, closer to dusk, thinning out a radish patch when I heard a rustle to my left. There was a gray fox not ten yards from me. We both froze for a second or two until it walked calmly away and hopped through an opening in the side of the barn. I followed and saw her inside, or maybe him, under some scrap wood with four kits.

I had warned the boys at least once a week that the chickens had to be locked up at night *because an animal would eat them.* Had I never considered the daylight hours in springtime, when foxes are trying to feed their hungry offspring? No, I had not. I left the ladies wandering the yard like a plump buffet, free-range and one hundred percent organic.

I cleaned up the girls as respectfully as one can with a shovel and a trash bag. I put the bag in the garbage can because I didn't want Emmett to see what was left of the bodies, and I put a rock on the can so the fox or a raccoon couldn't drag them out again. Then I went inside to tell my son about two more dead pets.

He was still up in his room; my initial self-loathing and disposal had required barely five minutes.

"Pup? I need you to come down for a minute."

I had no idea what to say. There was no snake vet here to bail me out. Louise wouldn't be home until after dark, after Emmett was supposed to put Comet and Snowball safely away for the night. I was flying solo.

Emmett appeared at the top of the stairs, paused, then came down slowly. He stopped on the landing three steps from the bottom. I was leaning against the wall. "You should have a seat," I said as evenly as I could.

"Am I in trouble?"

"What? No, you're not in trouble," I said, lowering myself to his level and reaching for his hands. But holy hell, that stung. I'd come to tell him I'd just let his chickens get killed, and he heard my comforting tone as menace? Did I sound *that* callous about death, if only out of habit? I didn't know that while I was cleaning up dead pets, he'd been ripping open the chocolate chips we were saving to bake cookies, but it wouldn't have mattered. Nobody was in trouble but me.

In through the nose, out through the mouth.

"We have a fox," I said. "And foxes are wild animals." Ease into it, cast the villain, set the stage, same way any good murder story gets set up. "And they, you know, hunt other animals. That's nature."

Stalling. He looked puzzled.

"I'm so sorry, pup, but he got the chickens." Rip the bandage off. But say *chickens*, impersonal, blunt the edge, if only for an instant.

"Comet and Snowball?" And the instant was gone. His eyes flooded with tears. "Are they dead?"

Now my eyes were wet. I couldn't get words around the rock in my throat. I nodded and pulled him close, and we stayed like that, softly crying, until I could swallow again and tell him what happened.

Emmett listened, moist-eyed and sniffling, while I recounted everything, from the blur by the steps to the dash into the underbrush, but without most of the gore. "It would have been quick," I said. "I'm sure they didn't suffer."

I wasn't sure of that at all.

He asked if he could see them, and I said that wasn't a good idea, that it was better for him to remember them as live chickens, the way they sat on his shoulder and jumped for blueberries. He asked if we could bury them next to Cosmo, and I told him that wasn't a good idea, either, because it would be awful if the fox or a stray dog dug them up. I didn't want to be aging them, like prosciutto.

He wanted to see where it happened. We went outside. Somehow I hadn't noticed the feathers scattered from the barn to the bushes, past the silver maple and Cosmo's grave, across the dirt patch we needed to reseed but didn't because Comet and Snowball ate whatever seeds we scattered. Emmett gathered up two big handfuls of feathers and walked to where we'd buried his pet snake less than a year earlier. He found another broken slate nearby, wedged it into the dirt like a tombstone, and piled what he had left of his chickens in front of it.

• • •

Why Peacocks?

I called animal control to let them know a chicken-killing fox was rampaging through my neighborhood. The lady who answered was not sympathetic. "That's what foxes do," she said. I told her I was plainly aware of that fact. She took my number, and a sheriff's deputy called back shortly after. He was very sympathetic. A chicken man himself. Laughed when I referred to the fox as "that little fucker," though he knew I wasn't making a joke. "I get it," he said. "But, you know, you do have the right to protect yourself and your property, and chickens are property. So you've got options."

I took his point, yet there were no options, not really. When the family of foxes had moved into the barn years earlier, I'd called a trapper to get rid of them. He gave me a price and we scheduled a time, and just because I was curious, I asked where he took them, if there was some forest preserve teeming with foxes hauled out of residential neighborhoods. "Uh, no," he said in a weary tone, as if he'd had this conversation before and knew how it was going to end. "We have to put 'em down. It's a state law."

I canceled our appointment, waited for the foxes to leave on their own, then bug-bombed the bejeezus out of the barn to get rid of the fleas. It wasn't their fault they were foxes. Maybe one of those kits, all grown up, had come back and killed our chickens. The lady was right: That's what foxes do. I'd just made it easier by leaving our two very trusting, utterly domesticated, and well-fed pets alone and out in the open for easy hunting. No, hunting would have been more of a challenge. Our yard was an abattoir.

Emmett took it upon himself to tell Calvin when he came home from school, and Calvin seemed rattled as much by the realization that toothy, bloody nature was occurring on the other side of the kitchen door as he was by the dead chickens. I didn't think he'd ever seen Comet and Snowball fight over a baby snake or peck a slug to death on the hot brick walkway. Funny how I never felt bad for the snakes or the slugs.

"They had such a happy chicken life," Louise said that night. The four of us were in the living room in what amounted to a subdued Irish wake. "I know they would have lived longer if we'd kept them locked up, but they wouldn't have been as happy."

There was no disagreement. We spent an hour or so telling funny chicken stories, which, when you think about it, is a tremendous amount of material for a pair of barred rock hens to generate.

"I hate to use the phrase 'a fate worse than death,'" Louise said, "but, really, can you imagine if one of them had survived by herself? How lonely she would be? They spent every moment together."

Yes, I thought but did not say, I could imagine that with great clarity. One of them would not be dead. We would believe that she was sad, which might be true or might be us projecting emotion onto a species that we otherwise slaughter by the tens of billions, that our own family eats in one form or another twice each week if not more. The surviving Comet or Snowball might be perfectly content with simply being not dead. I felt queasy again. "She would have been miserable," I agreed. But I wondered if a long life in a big pen was really worse than a short one picking bugs out of the grass.

An interesting thing about chicken wire is that it is meant to keep chickens contained rather than protected. That is, predators, including the foxes and raccoons miserably evident in the yard, can chew through it without much difficulty. Professionals know this, as would anyone remotely observant in the fencing aisle at Home Depot: The material we laymen call chicken wire is actually labeled with the far less robust title of *poultry netting*.

Why Peacocks?

I learned this the day after Comet and Snowball were killed because it occurred to me that the peacocks were vulnerable, too. They might be big enough and feisty enough to defend themselves during the day, but if some critter slipped in past midnight, when they were dozing and half-blind? Carnage. The pen, proudly assembled from leftovers and scraps, needed to be fortified with store-bought hardware.

My neighbor, the one with the goats and the miniature horse, offered to lend me the electric fence ribbon that he staked in the front yard now and again to let Chief graze outside the paddock. That seemed ideal. A strip across the bottom and another near the top would painfully but not lethally punish any animal that tried to eat my peacocks. Upon further consideration, though, it was apparent that I, too, would be painfully but not lethally punished every time I forgot to turn it off before I went into the pen, which would be more than once and, thus, too often. There was also the non-trivial possibility that one of the birds, likely Carl, would manage to take a peck at it, which could be lethal.

After a day of dithering, I decided to cover the pen with goat fencing, a heavy-gauge wire welded in two-inch-by-four-inch rectangles, held in place with screws and inch-wide washers. Anything big enough to chew through poultry netting, I hoped, would be too big to squeeze through the goat wire. It went up pretty quickly. I had the pen covered in about an hour, and then I went back with more screws and washers, probably more than I needed. I'd saved Carl twice, once (maybe) from an owl and once from a grommet. I wasn't losing him to a fox.

I put a final, excessive row of screws and washers at the very bottom, where the goat wire touched the dirt, then stood up and grabbed the wire with both hands. I tugged once, then a second time, harder, with what I estimated to be the force of a dozen especially savage and well-coordinated racoons. The wire barely moved. It was taut, secure, possibly impenetrable. I took a few steps back to admire my ingenuity, then heard the familiar rattle of Mr. Pickle vibrating his train.

"Yes," I said, shifting my focus from the wire to the inside of the pen, "you're very welcome." Except I couldn't see him. Sunlight sizzled from the goat wire and the washers and screws, a blinding pattern of long lines and big dots. The brilliance of the wire made everything behind it seem darker, screened in shadow.

I leaned my face against the cage, blinkered my eyes with my hands. Mr. Pickle was in full blossom in the center, and I could see Carl and Ethel off to one side. I stepped back again. The birds were only shadows behind bright lines and dots.

They were completely protected from predators, and yet I'd defeated the reason for having them at all. It's like the tree that falls in the forest: Is a peacock still magnificent if he can't be admired from outside the garbage coop?

I slept restlessly for a week after the chickens died, waking up convinced I heard the fox clawing at the peacock pen. But I never saw it again. Ethel honked only at squirrels and cats. The fox had cleaned out the easy treats and moved on.

One of us would find a barred feather every so often, stuck in a bush or hiding in the tall grass. The pile Emmett had laid in front of the tombstone lingered a surprisingly long time, a few of them pelted into the soil by hard rains. The last ones didn't blow away until the middle of April, by which time we'd all gotten used to not having to secure Comet and Snowball for the night.

Emmett didn't mention them again for several months, not until the end of June, Tater's birthday. Birthdays have always been an event at our house. There is a special crown and a banner and crepe ribbons strung in an exacting way from the dining room light to the doorjambs, and for the boys, Louise makes oversize and elaborate birthday cards covered with photographs from the previous year and a poem inside. They've saved every one. Tater's party was less fanciful, marked only with an extra treat and a new stuffed bear—he's a dog—and a tiramisu cake for the people. We sang "Happy Birthday," which seemed to both please and confuse him.

"Congratulations, Tater," Emmett said when the song was over. "You're the first pet that's made it a whole year."

Chapter Seventeen

Danny Potente slipped up beside me before I saw him coming, stood close, and spoke softly. "They in your blood?"

I gave him a quizzical look, fairly certain I'd never been asked that question before. We were standing near the doorway of a narrow beige-walled conference room on the first floor of the Four Points by Sheraton near the Kansas City airport. "I have no idea, Danny," I said. "What's supposed to be in my blood?"

Danny never broke eye contact. "C'mon, you know." He had a sly, vaguely conspiratorial smile. "The *peacocks*."

Over Danny's shoulder was a rack of sweatshirts and hoodies screened with the logo of the United Peafowl Association, the initials UPA in a tall, thin font with a peacock perched in the loop of the P, his train flowing past the base of the letters. Danny was the president of the UPA, which is sort of the peafowl equivalent of the American Kennel Club, except exponentially smaller.

"Um, no," I said. "Not in the blood."

"Oh, it'll happen. Just you wait." Danny was in his early sixties, with a mustache and swept-back black hair that was just beginning to thin. I'd called him a few weeks earlier but was still getting used to his accent, purebred New York, none of the gilded lilt one might expect for a fancier of extravagant birds. On the call, we'd patched in Loretta Smith, the vice president. With the two of them, I was trying to figure out how to salvage my peacock situation, three birds whom I couldn't let out but couldn't admire from the outside of their cage, either.

Neither one had much of an answer. Loretta paraphrased Winston Churchill, who paraphrased Sir John Lubbock, talking about a horse. "There's something about the outside of these birds," she said, "that's good for the inside of me." Well, yeah, that had been the original idea, before Carl got sick and the chickens got killed and I could see Mr. Pickle without going in the pen. But now what?

The UPA's annual convention was coming up, the twenty-fifth, so I signed up as a dues-paying member and booked a room at the Four Points. I'd never been to any kind of a convention, but they'd seemed like good places to learn things.

"How many you got again?" Danny asked.

"Three. Two boys and a girl."

"Oh, that's not good. You gotta get more hens."

Danny is a salesman. He's the CFO of a company he founded that designs and installs art and mirrors for hotels and hospitals—those pictures in all the rooms at the Radisson gotta come from somewhere—but he started as a salesman. Mass-produced art, mostly, and door-to-door from the trunk of his car. Old school. The key was getting whatever he was selling into the hands of the person he wanted to be a paying customer, have him or her hold it, trick the brain into a false sense of ownership. There must be a birthday coming up, he'd mention offhandedly, an anniversary maybe, and you know, this piece would look beautiful over that couch, don't you think?

Danny wasn't selling me birds, just the idea of them.

"I'm gonna take a hard pass on that, Danny. Three's enough. Maybe too many."

"No, no, no. *Three?* That's not too many."

"How many do you have?" Loretta had lowballed her birds at five hundred, but she was a breeder on a big spread in semi-rural Ohio. Danny was on Long Island.

"I'd have to count," he said. "A lot. I got all kinds. Ducks, chickens, pheasants, peacocks. I'm telling you, it's in my blood. My father had birds, his father had birds, *his* father had birds—"

"Wait, you don't know how many you have?" I said. This was suspect. "Where do you keep so many birds you can't even count them?"

Danny looked in the lobby and glanced back into the room, as if checking to see who was in earshot. "I free-range most of 'em," he said. "I got places."

"Places?"

"Yeah, you know, places. If one place becomes a problem, I go to another place. I know people."

"A problem?" I'd stumbled into the underbelly of the peacock world.

"Yeah, you know, sometimes there can be problems and—You know what? Forget about it. Doesn't matter." He did another head check, which I realized was less about who might be listening than whom he should be greeting, being the president and all. "You need more birds," he said.

"No. I really don't."

"Heh. You wait. This time next year, you'll have fifty. I'm telling you, they get in your blood."

• • •

The United Peafowl Association, not surprisingly, is a niche organization. At the time, which was 2018, there were 223 members scattered from Florida to Texas to North Dakota, plus eight more overseas in places like Belgium and Thailand. For an outfit dedicated to the world's most ostentatious bird, there is very little glamour and no pretension; the UPA is like Linus's pumpkin patch, all sincerity. Even the milestone twenty-fifth annual convention was an unassuming, almost cozy affair: There were only thirty or so of us in attendance, and the banquet dinner was the early bird at a Golden Corral near the Four Points.

The members who traveled to Kansas City were a mix of hobbyists like me, serious breeders of varying degree, and what would fairly be called enthusiasts, people who simply like peacocks and peacock-related things, of which there is an astonishing assortment. The peacock print and the peacock shape and the peacock's colors can be applied to most consumer products and solid surfaces. The conference room at the Four Points was ringed with folding tables covered in black cloth displaying items to be auctioned later. There were peacock umbrellas and place mats and napkins and bookends and notebooks and wrapping paper and wind chimes and brooches and a yellow diamond-shaped sign that said "Peacock Crossing" and pencil holders and ornaments and many framed prints.

I spent the better part of a half hour taking it all in, and thought of a woman I'd met in California, where the peacock killer was still at large. Cat Spydell took in abandoned and wounded animals on what she called the Pixie Dust Ranch but was really just the yard sloping down from her house in Palos Verdes. One of her rescues is a peacock she raised from a chick. She named him Radagast, after the Tolkien character, and Rad for short, and she took him to schools and festivals and such, like an educational ambassador on behalf of peacocks everywhere. One of the side benefits of having a peacock, Cat said, is that he was more glamorous than Lily the potbellied pig, who used to be her most famous rescue. "You know what happened? Everyone gave me pig things," she told me. "Pig earrings and pig mugs and pig key chains—they're all *pigs*." She let the pig-ness settle for a moment. "Now people give me peacock things. I don't think you can have too many peacock things."

I don't know, I thought, looking at the peacock necktie between the peacock wind chimes and the peacock earrings. *There might be a limit.*

But I wasn't there for the tchotchkes. I was there to learn, to gather practical knowledge from people who actually understood these birds. The morning of the first full day, right after breakfast, there was a seminar on how to examine fecal samples for parasites, with both microscopes and poop provided, so it was hands-on. That would have been a necessary and unpleasant skill if I didn't have Burkett nearby. In the afternoon, a breeder gave a presentation on green peacocks, which to some people's taste are prettier, more exotic, than the India blues. Because greens are so cold-sensitive, though, they're often bred with blues to put some of the former's tropical elegance into the latter's climatic sturdiness. The hybrid is called a Spalding, named for the woman who first crossed them.

"Who's that?" I whispered to Danny.

"Who? Spalding? The woman who first bred them. He just said that."

"Yeah, but who was she?"

Danny shrugged. So did everyone else I asked.

Danny was supposed to give a presentation on how to free-range birds. His places, it turned out, are all legitimate—the rail-road tracks behind his office, a restaurant that invited him to leave some birds, an industrial area that doesn't mind, a big fenced lot the owners said he could use, a friend's aviaries. His accent just made it sound sketchy, like something out of a weirder Scorsese movie. But he got distracted by feral cats. "Killing machines," he calls them. Millions of cats kill billions of birds every year. He went on like that for a while and never did get back to free-ranging.

That evening, Twain Lockhart led his presentation on peafowl nutrition with a blank slide. Twain, a poultry consultant at the big feed company Nutrena, had a bushy beard and a blue ball cap, and I believed it was entirely possible he'd driven a tractor to Kansas City. He also brought a bag of little candy bars to toss out, ostensibly as a reward for answering questions, but the whole thing was more randomly jolly than that. With the blank slide behind him, he spread his arms for effect. "Here's all the data we have on peafowl, folks," which clearly was none. Twain was there to explain why there is no dedicated peafowl feed, which was a matter of basic economics. At any given time, there are approximately twenty-three billion chickens on the planet, almost all of which are white leghorns or a close relation thereof, laying in the commercial egg industry, or varieties of Cornish Cross living short, horrid lives on commercial farms. "Eight weeks, beginning to end," Twain said. "Fifty-six days from hatch to the freezer."

Comet and Snowball had a hell of a life.

Those billions of birds are the ones the feed companies care about, the ones making up ninety-five-plus percent of the market. Turkeys, forty-six million of which are eaten at Thanksgiving alone, get their own feed, and there are enough pheasants raised for hunting and eating and showing to justify a specialty feed. Peacocks are counted only by the thousands, or they would be if anyone took the trouble to count. "You've got a ways to go before the feed companies are gonna take an interest in peafowl," Twain said. "The PhDs [who formulate the feeds] say they're just fancy chickens. Big fancy chickens."

A few heads shook slowly, and there was a general titter of disappointment.

Most people are familiar with only one kind of peacock, the India blue. Maybe they've seen a white one, but that was probably a leucistic India blue. In any case, the peacock's palette, to the layperson or even the owner of three birds, is somewhat limited.

That apparently was a failure of my imagination: The UPA recognizes 225 varieties of peafowl. Not species—there are still only three, blue, green, and Congo; no one is reordering the animal kingdom—but genetic mutants that have been crossbred and rebred to create thirteen colors and five patterns, such as silver pied and white-eyed, which is when the eyespots in the feathers are white. And that's just by UPA's current count. In 2017, *seven* new colors were presented for the UPA's consideration by Legg's Peafowl Farm in Kansas City. It's about thirteen miles from the Four Points and the main reason the convention was in Missouri: Brad Legg, who's been in the business more than fifty years and is pretty much a legend in the field, was opening up his farm for a day, and peafowl fanciers wanted to see the new colors and all of the old ones, too.

Old is a relative term. The first three mutations—black shoulder, white, and pied—were all documented in the eighteen hundreds. (They surely had appeared before then, but nobody wrote it down.) The next variety didn't show up until the late nineteen sixties, a brownish bird hatched in Maine and originally referred to as a silver dun before the name was changed to cameo. The tricky part isn't finding mutants so much as getting them to breed true—that is, making sure a bird with a silver dun mutation in its genetic code is capable of producing new silver dun chicks. That process typically involves generations of deliberate breeding, much of it with mothers and sons and fathers and daughters but not so much that the offspring end up hobbled by that same inbreeding.

Twenty years after the cameo was stabilized, breeders started locking down other mutations. Charcoal, bronze, and white-eyed were all introduced in the eighties. The first purple was hatched on an Arizona farm in 1987, and opal and silver pied were bred in the nineties. Peach came from a multigenerational breeding of purples and cameos. Brad Legg found a midnight bird at an exotic bird sale in 1998, jade at another sale in 2000, hatched taupe from a purple hen and a plain India blue male in 2005, and found a steel bird at an animal swap meet in 2010.

Once the patterns are crossed with the colors, the combinations multiply exponentially. The genetic leaps are relatively short from an opal pied to an opal black-shoulder pied to an opal black-shoulder pied white-eyed.

Why anyone would want to make those genetic leaps at first struck me as odd, borderline obsessive, even. Peacock and peahen breeding is a precision exercise at this level, requiring copious and detailed records of which eggs came from which cock and hen combo and which eggs those cocks and hens hatched from to begin with, and so on back through the generations. Yet there will never be a substantial market for opal black-shoulder pied white-eyed peacocks. There are only so many people with the space and the patience for birds whose sole purpose is to be pretty, and fewer still who want an even more exotic version.

But so what? There are enough collectors, enough fellow breeders. And how is it any different than, say, breeding dogs for certain characteristics, like bug eyes and smashed face? The genetic gap between Tater and a German shepherd is vanishingly small, too.

And to be the first person to breed that opal black-shoulder pied white-eye? In the peacock world, that's a triumphant achievement. It doesn't matter if nonenthusiasts don't get it. Journalists give each other awards all the time that mean absolutely nothing to anyone outside the business. So do Realtors and accountants. I've got a friend who spends his weekends racing junker cars and another who hunts mushrooms. Everyone's got their thing.

Opals and cameos were not my thing.

"Not a lot of blue here, Danny," I said. It was damp, overcast, the ground muddy from a week of rain. We were standing next to a pen of birds of a new color, Montana. Legg gave them that name because he drove all the way to Montana to get the first one and because it was the color of high-plains grasses.

"Yeah, but that's a beautiful bird," Danny said.

"It is, it is. It just doesn't look like a peacock. Most of these colors aren't really, you know, *colors*."

"Get outta here. Of course they're colors. Did you see the platinum?"

"I did, I did. Gorgeous. Not a peacock color."

He gave me a sideways look. "You don't like the pastels?"

"Is that what you call them? Not sure I'd call ivory or mocha a pastel. And charcoal? That's just a black-and-white bird. Like all the peacock got bred out of it except for the feathers."

"Oh, c'mon. These are gorgeous."

I thought just then of a line from a Flannery O'Connor novella called "The Displaced Person," wherein a peacock is a sort of moral litmus test. One either saw in a peacock "a tail full of suns," she wrote, or one was Mrs. Shortley, who saw "nothing but a peachicken."

That's why Danny was the president of the UPA and why I had three peacocks hidden away in a garbage cage. Compared to him, I was Mrs. Shortley. He saw suns I could not. The birds were in his blood, he'd said, and I understood what he meant.

I looked down a long stretch of runs and caught not a single glint of blue. But no one else seemed to mind. They were enthralled with these new mutations, redecorated and toned down by very determined and patient mortals. A few varieties, like steel, appeared more sophisticated, reserved, as if perhaps they were dressed for court. The rest, to my eye, looked either washed out or too dark. But maybe there was something I couldn't see. I was clearly in a minority. Peacock aficionados appreciate a finely crafted novelty.

I remembered something Twain Lockhart had said the night before. The peacock industry would need another fifty years or so to catch up with pheasants and build a market large enough to get the attention of the feed companies. But how? Pheasants are bred for commercially viable reasons, mostly to be hunted or slaughtered, but reasons nonetheless. Peacocks are bred to be looked at, but only in limited circumstances. There are no peacock shows, no peacock competitions. There's no market to develop because there aren't enough people with the interest, space, and tolerance for noise and poop to sustain anything more than a niche.

A Montana peacock, I told Danny, will never be written into a fairy tale.

"So what?" Danny said. "Still a gorgeous bird. And really, isn't this something?"

*   *   *

Why Peacocks?

On the last night of the convention, I met a man named Ray Watts from Macon, Georgia. We were in the bar of the Four Points, Ray and I and an Oklahoman named Mark with big hands and many stories of turkeys and loose dogs. I told them about Carl getting sick and how Burkett had told me to sneak up on him in the dark on a ladder.

"I use a catch net myself," Ray said. I made a mental note to thank Uncle John again. "And I don't like to go in the pen at night because I'm afraid of snakes." I made a note to check for snakes in the dark.

Mark told me he was a retired firefighter, and I told him I'd written a few stories about firefighters over the years, and our conversation drifted far from peacocks. Ray sat quietly, as if concentrating on something important.

"I sold some birds to another writer once," he said finally, partly to us and partly, it seemed, to jog his memory.

"Really?" I said. "Anyone I'd know?"

"Maybe. Kinda famous," Ray said. "Funny name. Real funny name. Up in Milledgeville."

I was lifting a beer but stopped, set it down gently, peered at Ray. "Any chance," I said, "the name was Flannery O'Connor?"

Ray's face lit up, and he slapped the table. "That's her!" He settled back, triumphant. "She was dead, of course. But I sold a pair of peafowl to the place she used to live. It's a museum now."

Louise and I kept planning to go to Andalusia, the O'Connor family's farm that's now a museum. We hadn't made it, though, not yet. Life gets busy, and slipping away for a weekend in Georgia is harder than when we had one cat and no kids. Besides, there weren't many peacocks to see there anymore, just the ones Ray delivered. We had our own birds to keep us occupied.

When O'Connor was alive, dozens of peacocks roosted in the trees and on the buildings and the sagging gateposts. She decorated letters and gifts with feathers, which she also gave to the ladies of Milledgeville to put in their hats. She tucked the birds into her stories as scenery and symbols and acid tests of character, like she did in "The Displaced Person." O'Connor and peacocks are so iconically entwined that the cover of one of the paperback editions of *A Good Man Is Hard to Find* is a drawing of a somber, hollow-eyed woman in a long skirt and black hat with a fan of blocky peacock feathers arranged on her backside.

O'Connor bought her first peacocks, a mating pair and four chicks, in 1952, and by the time she got around to writing about them nine years later, she had at least forty at Andalusia. "The population figure I give out is forty, but for some time now I have not felt it wise to take a census," she wrote in a 1961 essay for *Holiday* magazine that was later republished as "The King of the Birds" in an anthology of her work. "I had been told before I bought my birds that peafowl are difficult to raise. It is not so, alas."

She, too, began with chickens, when she was five years old, in particular a buff Cochin bantam that walked backwards. Pathé News caught wind of her chicken's curious ability and sent a cameraman to film a short newsreel. "Here's Mary O'Connor"— Flannery was her middle name—"of Savannah, Georgia, holding the only chicken in the world that actually walks backwards," the narrator said in that nasally patter of thirties hucksters. "When she advances, she retreats; to go forward, she goes back," and so on.

It was a bit dubious. Surely other chickens were capable of walking backward. But O'Connor's was the one that went Depression-era viral, and after that minor burst of fame, she began collecting all manner of fowl: pheasants and turkeys and swans and ducks and quail. "My quest, whatever it was actually for, ended with peacocks. Instinct, not knowledge, led me to them," she wrote, which wasn't so different from how Mr. Pickle and Carl and Ethel ended up in my yard. "I knew that the peacock had been the bird of Hera, the wife of Zeus," she continued, "but since that time it had probably come down in the world—the Florida *Market Bulletin* advertised three-year-old peafowl at sixty-five dollars a pair."

That is the most concise and accurate history of the peacock ever written.

Chapter Eighteen

I bought a peahen from nurse Valerie.

This was not something I did with enthusiasm. Peacocks were not in my blood. I was not eager to expand our menagerie of ornamental, and unexpectedly expensive, birds. It was a concession to reality, a simple risk analysis of the looming breeding season. The confounding part was that I didn't actually want the birds to breed. I wasn't running a hatchery, and I had neither the time nor the space for the half-dozen chicks Ethel might produce. But nature doesn't much care what I think. Peacocks evolved those magnificent trains for the sole purpose of convincing peahens to copulate; the display exists in service of breeding, and one can't marvel at the former without tolerating the latter. Buy the ticket, take the ride, as they say.

Because breeding is at root a competition, peacocks sometimes fight. Not viciously, but Carl and Mr. Pickle would be better off separated for a while. With only a single hen, one of them would be isolated. That would be young Carl, who I was sure was not yet capable of breeding. He insisted otherwise, throwing up his feathers with a fierce urgency since his return, pawing the sand like a two-footed bull. But it was a desperate, overwrought imitation of sexual maturity. He still had only the beginnings of a real train, most of it the barred feathers of a juvenile that, instead of creating a smooth arc, jabbed upward in random, spiky lengths that made it look like something he forgot to comb. A half-dozen eyespots had come in, but all on the left. Even at his most beautiful, he was lopsided.

Was alone the same as lonely for a peacock? Would being physically separated from Mr. Pickle and Ethel have an emotional effect on Carl, and would that emotional effect manifest as a physical one? Was I overthinking the social and mental health of a bird?

Yes, I was definitely doing that. But Carl was an investment now, and investments had to be protected. I couldn't risk another injury, a bloody wound in a sex fight with Mr. Pickle or a psychic trauma from solitary confinement. Burkett didn't cleanse that bird of three toxic metals just so Carl could mope himself to death.

So I needed a second hen to keep company with Carl.

Louise was not thrilled with the idea of adding to the flock. "We're up to *four*," she said. "More than that and you're a hoarder." The drama and expense of Carl's illness was fresh on her mind, though, and the thought of Carl landing in Burkett's operating room again was, I suppose, slightly worse than the additional honking and pooping of one more bird.

I never told her how much three weeks in an avian hospital and at least nine hours of surgery had cost because it was embarrassing. That was the exact answer I gave her when she asked for a number: "Embarrassing." Large enough that it wasn't rude to negotiate. Not used-car-level negotiating, maybe, but that is not an imperfect analogy. I'd recently come into an unexpected lump of cash that covered the bill, so the financial hit was tolerable. Yet it had been a point of scruffy pride that my mythical birds were purchased for a flash-sale price and housed in scraps and trash that, if we'd ever gotten around to it, we would have had to pay someone to haul away. In the abstract, I could lie to myself, they were *saving* us money, using up all those old boards and rolls of rusty wire. And that was part of the wonder of having peacocks, that they were so cost-effective. The fantastical beauty of peacocks for less than the price of chickens! Until Carl ate a grommet. My birds now were precisely what they appeared to be, a lavish, impetuous indulgence.

Louise did not push me for a number. She knew I wasn't hiding it so much as I just didn't want to say it out loud (or, for that matter, type it in a book). And I knew she had agreed to only two birds to begin with, not three.

"Okay, okay," she said, letting me off the hook. Then she narrowed her eyes a bit and her crafty smile emerged. "In a marriage, each person gets one secret," she said. "And you just used yours on Carl."

"Wait a minute! I didn't know we got to have a secret."

"Too late! Don't tell me." She put up a hand.

Didn't matter. I'm terrible at secrets. Too neurotic. Had she pressed me for the vet bill, I would have produced it in shame, which she knew full well. "Well, what's yours? We can just say them now, get it over with."

"Can't," she said with a shrug. "Don't know yet. But I'm gonna live to be ninety. I've got lots of time."

I told her Valerie wanted fifty bucks for the new hen. She agreed that was a tolerable number to keep Carl, now our investment bird, safe.

One of the few things I'd learned for certain from direct observation was that peacocks are social creatures. Ethel often roosted next to one of the boys—she didn't seem to have a preference—and they moved as a loose group when I was in the pen. They appeared to take cues from each other: The day after Mr. Pickle snatched a blueberry from my fingers, Ethel ate them from my palm, and Carl took them out of my hand his first day back. Ever since, they'd lined up in front of me, side by side.

Still, I had to ease the reintegration. Before I brought Carl home from the hospital, I blocked off a section of the pen with a panel of chicken wire, making a kind of penalty box. It might not have been necessary, as Ethel trilled and cooed to him through the wire and Mr. Pickle mostly ignored him. When I let him out on the third day, I called Valerie to thank her for the tip about the sand and ask if she knew where I could get another hen.

"Craigslist," she said. There was a pause. "You know what? I've got an extra girl you can have. She's a Spalding. Fifty dollars all right?"

I had no idea if a Spalding was the kind of hen we should have. Or how much one should cost. But this is apparently how I buy my peacocks—with as little information up front as possible.

The cross-bred product of green peacocks and India blues sounds exotic, but Spaldings are actually quite common these days, having been bred and rebred since the first half of the twentieth century. I mean, hell, I bought one for fifty bucks in North Carolina. Of the United Peafowl Association's 225 approved varieties, 111 are Spaldings; for every cameo black-shoulder pied white-eyed, there is a *Spalding* cameo black-shoulder pied white-eyed.

The hybrid is called a Spalding after the person who successfully developed the breed, who, since at least a 1959 issue of *Modern Game Breeding and Hunting Club News,* has routinely been identified only as "the late Mrs. Keith Spalding in California." Occasionally, her husband will get clipped and she'll be referred to as Mrs. Spalding, and sometimes she's acknowledged to have been a bird fancier, which would seem self-evident. But for a woman who conjured a bird, she's conspicuously cast as a mere footnote in peacock history.

Before she was Mrs. Spalding, she was Eudora Hull Gaylord, a wealthy and public-spirited widow from Chicago. In 1905, in memory of her dead husband, Edward Gaylord, she founded and funded the Edward Sanatorium, an open-air sixteen-bed center for the treatment of tuberculosis, which at that time killed about four thousand people every year in Chicago, including her husband. "One of the most complete tuberculosis camps in the country," a San Francisco newspaper called it, "the proposed camp is to be used for treatment of the poor only. Only incipient and curable cases will be received."

She was thirty-nine, a full decade older than Keith Spalding, when they married in 1906. He was another wealthy Chicago native, a manufacturer of steel products, and the son of Albert G. Spalding, founder of the eponymous sporting goods empire and, before he retired from the Boston Red Stockings at the age of twenty-six, the first pitcher to win two hundred games. (Albert was also president of the Chicago White Stockings, now the Cubs, from 1882 to 1891, during which time the team won three pennants; and he was one of the people who made up the story about Abner Doubleday having invented baseball in Cooperstown, New York. Albert is in the Hall of Fame, of course.)

The two of them, Eudora and Keith, spent much of their time in California, where she had fifteen hundred acres of citrus groves and lima bean fields in Ventura County. Eudora's father had bought what was left of a Mexican land grant called Rancho Sespe from the heirs of T. W. More, who was shot to death squabbling over the boundaries and water rights. Eudora's father, in turn, left it to his two children, and Eudora bought out her brother, Morton, who later would spend a decade in Congress representing the second district of Illinois. There was a bunkhouse for single white males designed by the architects Greene and Greene, who were pioneers in the American arts and crafts movement, and separate quarters for the Japanese workers. The Mexican laborers, meanwhile, lived in a village on the ranch where they had plots to grow their own vegetables. By 1910, Rancho Sespe was one of the largest lemon groves in the world.

Eudora and Keith had a yacht built in a Delaware shipyard, a 161-foot schooner called *Goodwill,* that Eudora, cradling a spray of roses, christened in 1922. *Goodwill* crossed the Atlantic and sailed the South Pacific and had a launch from which Keith and Eudora angled for big fish, a sport at which she excelled. She caught a marlin big enough to make the newspaper in 1937, and that wasn't even her biggest catch: She once reeled in a monster broadbill that weighed 426 pounds. Zane Grey, the famous author of Western novels, talked smack about her, arguing months after the fact that she couldn't possibly have the strength to land such a huge fish. Jealousy, plain and simple.

There are no public accounts of Eudora and peacocks, let alone any acknowledgment that she created a successful hybrid. But she was indeed a bird fancier. Mrs. Keith Spalding is mentioned, again in passing, in at least one brief history of pheasant breeding, and *The Catalina Islander* in May 1933 noted her tour of the island's Bird Park "in which she is keenly interested . . . Both Mr. and Mrs. Spalding consider the Bird Park the finest and best arranged it has ever been their pleasure to visit." In 1929, she and Keith donated the first California condor to the San Diego Zoo after it showed up at Rancho Sespe with a crippled wing. Two years after that, she bought a pair of cassowaries that she kept in pens fifty feet wide and a hundred feet long, which is something only a hardcore fancier of birds would do because a cassowary is like an ornery emu with a helmeted blue head that might try to kill you. And no later than 1936, Rancho Sespe Game Farm was advertising five varieties of peafowl for sale in *The Game Breeder and Sportsman*: "Blackshoulder, Green, White, Blue and Spalding."

When she died in 1942, Eudora put the ranch in a trust to benefit the California Institute of Technology. Keith administered it, and gave Caltech money for the Eudora Hull Spalding Laboratory of Engineering, which was built on its Pasadena campus in 1957. It is the only building there named solely for a woman, and an oil portrait of her with brown curls and brown eyes and a Mona Lisa smile hangs in the foyer. There is no mention in the placard below the painting of her hospital or the ranch or the huge fish she reeled in or the namesake birds she created.

Our new hen's auspicious providence helped ease any tension that remained between me and Louise over my growing flock. There's nothing Louise loves more than a fierce old lady with bold ambitions and eccentric hobbies. Had I done my research before committing to another hen, I would have led with Eudora.

Valerie lives about ten miles west of us, on the other side of the city, in a low brick house on a corner lot surrounded by chain-link fencing. I took the boys with me on a Sunday afternoon in late March, and we waited in the car while she pushed a button to swing open a wide gate. There was a sign warning us to watch for chickens.

Valerie waved to us from a wooden deck attached to a three-season porch and next to an aboveground swimming pool that hadn't been opened for the season yet. There were a couple of garden beds in the middle of the yard, and the lawn followed a shallow slope down into tree shade at the back of the lot. It was a fairly ordinary semi-suburban backyard, except for all the birds.

There were too many to quickly count, which would have been difficult anyway because almost all of them were roaming free, a few of them madly scurrying, disappearing behind plants and tree trunks. There were chickens, big ones and medium ones and a swarm of miniature ones called Sebrights that had copper feathers laced with black, as if each one had been outlined with a thick felt pen. There were ducks, Indian runners and what looked like mallards and a white one with a poof on his head like a large cotton ball and another, bigger white one with a plain, regular head. Farther down the lot, under the trees, were coops and two pens made from fence panels, in one of which was a pied peacock and four hens. The yard wasn't crowded, though, as if it had been overrun with feral poultry. It felt more like a sanctuary, which in many ways it was, for both the birds and Valerie.

"I've read some of your stuff," she told me once. "Our stories aren't so different."

What she meant was that we both made a living amid misery and death, the real discrepancy being that her job was useful and noble, as opposed to mine, which I'd come to suspect was opportunistic and parasitic. Valerie for many years was a nurse in an intensive care unit and a neonatal ICU. Burnout is distressingly high for ICU nurses—in studies, up to eighty percent report suffering some symptoms—because the job is exhausting, physically and emotionally. Valerie retired young. By the time I met her, she managed a portfolio of rental property that had been in her family for years and was raising her son alone after a divorce. Graham is a quiet, bright kid with Asperger's and, as Valerie describes him, sort of a bird empath. While he can find humans and their complicated social cues frustrating, birds for him are soothing, intuitive, simpatico.

This started to become apparent in the early years of elementary school, when Graham brought the class finch home over holidays and long weekends. In second grade, he asked for a pair of birds for Christmas, and Valerie got him two zebra finches, tiny things too small to bite hard and too even-tempered to bother. She thought she'd bought a bonded pair of males that Graham named Tweet and Sweetheart, but not three months later, Sweetheart laid an egg. So Graham and his mom were now breeding finches.

There is always the potential when dabbling with birds—and this no one tells you beforehand—of becoming enchanted, and it is impossible to understand this until it happens. Until, for example, two plump hens sprint toward the sound of your voice and Comet hops on your shoulder and Snowball leaps for blueberries and Mr. Pickle twirls so closely to you that his ruffling feathers brush across your face and Ethel tilts her head and softly trills and waits for another piece of tomato. It is utterly unexpected, too, because the experience is not at all like that of other pets. Tater is my constant companion, unshakably loyal, tirelessly available. Yet a dog withholds nothing, so there are no secrets. Birds, if we aren't paying attention, are all secrets, so alien from us that it hardly seems worth the effort to decode them.

But then there is, sometimes for some people, a starburst of insight, not of understanding but of realizing there is something to be understood. One bird will reveal one tiny secret—*I recognize you,* perhaps most often—and your perception of the natural world shifts. Those secrets become mysteries, which are different because mysteries are an invitation to explore, and each one reveals another and then those birds have personalities and intelligence and foibles and charms and *souls* and it all sounds ridiculous but it's true.

That's basically what happened to Valerie.

The finches led to parrotlets, which, as the name suggests, are itty-bitty parrots. Then she came across a plea on Craigslist from a woman who was moving and had to find a home for two ducks, two chicken hens, and a black Australorp rooster. *Help,* the ad said. *I need someone who will love them and not eat them.*

"Graham," Valerie told her son, "we need these."

Peacocks came next, also from a Craigslist ad, then more ducks, more chickens. She bought some, but most were rescues in one way or another—hens that had stopped laying, the rooster the 4-H kid was supposed to cull from his flock but didn't have the heart to kill, wounded ducks picked up by the ASPCA. She bartered her nursing skills for veterinary care from Burkett, which she was doing the day I met her, and she'd set up a kind of avian ICU in her three-season porch where she can administer subcutaneous fluids and tube-feed and such.

The white duck with the plain head in the yard was a Pekin named Daisy. She was the matriarch of that five-bird variety pack Valerie found on Craigslist, and she got horribly sick. Graham picked up on it. She'd swallowed a piece of wire. It had to be cut out, but it was deep in the thicket of her digestive tract, "and once you start opening up runs of bowel," Valerie said, "she's already gone." Burkett used X-rays from multiple angles to triangulate where to cut. Though he missed by only an eighth of an inch, the surgery took hours. When Daisy was sewn up and the anesthesia began to fade, "she stood up like an Indian runner duck and plastered herself to me."

(That analogy is more endearing once you understand that Indian runners hold themselves erect, like penguins, and instead of waddling, they, well, run. But it's pretty charming in any case simply for involving a duck.)

Valerie was explaining her fondness for the birds when she told me that story, telling me how they are always there, affectionate in their way, appreciative, never arguing or complaining, and she started to cry when she got to Daisy. She caught herself, laughed maniacally, and said, "I spent six thousand dollars to save a fucking duck."

I did not consider that outlandish.

On that Sunday in March, though, the boys and I were only picking up a Spalding hen. We'd brought a cat carrier with us, the one we used to take Okra to the vet. I'd learned my lesson about feed bags.

Valerie walked us down to the pen, pointed out our hen. She was, for a girl peacock, striking, as if she'd been run through a filter, the color saturation and contrast ratcheted up. Instead of pond-water brown, her back was a rich chocolate, the kind Whole Foods stocks in the checkout line. The green in her neck, which was longer and narrower than Ethel's, was closer to emerald than grass, and there was blue in there, too, just a shade duller and darker than Mr. Pickle's breast. A lighter blue, the color of an April sky, appeared to have been dusted below and behind each eye, and a slash of brown curled across the top of each, like drawn-on eyebrows. Her crest was noticeably taller and tighter. She was a glamorous version of Ethel, Ginger to her Mary Ann.

Valerie went in and got her, calmed her down, slipped her into the crate. "She'll be good with Carl," Valerie said. "What are you going to call her?"

"Girl Carl!" Emmett chirped.

"GC?" Valerie offered. "Abbreviate it?"

We settled on Carlotta once we got her home, though Louise continued to refer to her as "Carl's concubine." Later, we realized we should have named her Eudora in honor of the late Mrs. Keith Spalding of California—which also would have created the buddy-comedy title of Louise's dreams, "Ethel & Eudora"—but Carlotta already had stuck. As with children, there are no second chances on naming a bird.

Chapter Nineteen

The noise started in the middle of April, the beginning of mating season. Mr. Pickle, a rising two-note burst, E above middle C, up to G, a quick slur down to F-sharp. He repeated it twice, which I could hear from inside the house. It was not a plaintive cry, desperate and whiny, but assertive, a robust announcement: *I am here.* A moment later, he encored with a triplet of single notes in the same range, *mow mow mow.*

There was an immediate sense of relief. *That wasn't so bad,* I thought. He did not sound like a woman crying for help nor, I presumed, a dying child, which were all descriptions I'd read. He was not shrill. He made a luxurious and exotic noise, trembling with tropical steam and far-flung passion. It was a lithe and sultry sound, if one was determined to hear it as such.

For months I had dreaded those noises. I had clicked enough YouTube links, heard enough of the cries through the tinny speakers of a laptop to fear them. Not for us so much as the people who lived around us. I didn't want to start a bird war in the neighborhood, touch off another Palos Verdes. As near as I could discern, there was no city or county ordinance specifically outlawing peacocks, but in case one existed, I didn't want neighbors' complaints to draw attention to my flock. Ideally I would have considered all of this before impulsively bringing three birds home to satisfy a sudden, irrational, and most likely passing obsession. But it was too late. I was four birds in. I'd paid Burkett a small fortune and hoarded sand. More to the point, I liked them. Not all birds, necessarily, but these specific ones, Carl and Mr. Pickle and Ethel. I was still warming to Carlotta, but I assumed I'd grow fond of her, too.

I asked my shrink friend once why I didn't just euthanize Carl when he was sicker than Keith Richards. "Bonded empathy," he said. "He's your buddy."

"Meh," I said. "I think I was just too chickenshit to put him out of his misery."

He shrugged. "That's in Buddyville. Same thing."

It was all irrelevant, because none of the birds had yet to show the slightest inclination to leave. Sometimes I would test them, stand with the door open and wait for them to make a move. They always did, but only toward the cinder block from which treats were dispensed.

But the matter of noise remained. For a long stretch of delusional hope, I thought perhaps I would be immune, that maybe captive peacocks were quieter, even silent. Or maybe it wasn't so bad to begin with; peacock racket wouldn't have been the only phenomenon exaggerated on the Internet. Maybe some people even enjoyed the whoops and hoots and yawps.

That was, I briefly believed, entirely possible. On the first day of April, Uncle John, my friend with the net, sent me a posting from his neighborhood listserv. "Welcome the peacocks to Forest Hills!" it said across the top. A woman who lived near him explained that she had been on a meditation retreat in an old monastery near Rome. "A flock of peacocks roamed the grounds," she wrote, "and they made the experience of staying there absolutely magical. The adult males are the most beautiful birds on the planet, parading around with their gorgeous tails. In the evening it was fascinating to watch the hens herd their chicks onto the high perches, where they snuggled under their mommies' wings for the night."

All of that sounded about right. She was going to start her own flock and invited her neighbors to do so as well. Forest Hills is a historic neighborhood nearby, built around a long-ago-decommissioned golf course. So many places for birds to roam and mingle! This was a compatriot. This was also, I thought, a possible way out of my dilemma to eventually free-range my birds.

"And the best part," she wrote, "was waking in the morning to their soothing calls."

Okay, that was a little off. I'd never seen that adjective attached to a peacock, but I figured that she'd probably meditated in silence for so long that any ambient sound was soothing. I asked John for an introduction, which he graciously provided.

I missed the link at the end of the message. If I'd clicked, I would have heard another of those shrieking clips, quite possibly the least soothing one on the Internet.

All in all, a pretty good April Fools' Day joke for a listserv.

Had I'd become so earnest about my birds that I could no longer detect sarcasm? Yes, I had, and this only prepared me to be mortified by the coming mating season. My neighbors would be furious, and Louise would shake her head in quiet disappointment. The boys might complain, too, though they were more of a wild card. The chickens and the goats and Chief probably wouldn't mind, but they didn't get a vote.

Then, one fine day, Mr. Pickle started calling and it wasn't so bad. Not to my ear.

He did it twice more before sundown and once after and maybe more in the middle of the night, but if he did, I slept through it. That was a good sign. Carl joined in the next day. They made low, whistling hoots and a midrange *yow* and a higher-toned *ah ah ah ah*. The noises were occasional in April, then regular by May, at which point they escalated quickly to frequent and, at times, incessant. The squeak of the screen door would set them off, as would a car rolling over the gravel or the beep of a delivery van backing up.

None of it bothered me, but I accepted that my judgment was skewed. After Carl's extended flirtation with death, the boys had again lost interest, and Louise had her own deadlines to contend with. The peacocks were no longer ours but mine. Which was fine because they were minimal effort, just keeping them fed and watered and, every other day, scooping up droppings as if cleaning a giant litter box, and they were pleasant, at times even delightful, company. But it did make me predisposed to minimize their less appealing characteristics, so I relied on Louise as a more realistic barometer. If she referred to "the peacocks," that meant they'd been benign, and if she used a name, Carl or Ethel, there had been a pleasant interaction of some sort. "The birds" indicated mild annoyance, while "*those* birds" suggested a more agitated state. There were fine gradations from there. We hit "motherfuckers" only twice that summer.

Yet no one complained. I expected a knock on the door from a city official or a peeved, sleepy-eyed stranger, but it never came. One passing dog-walker asked if we had peacocks because she recognized the noise, but she did not seem perturbed. I began to believe that the acoustics of the pen, with a metal roof and my double-safe fortress wire, redirected the sound toward the ground, maybe smothered it. But then Emmett reported that, no, he'd been on his bike two blocks away and heard them quite clearly.

Then they stopped. Abruptly, as if one hot day in August they decided that their vocal cords were exhausted. Their feathers had begun falling out by that point, too, a correlation Pliny the Elder had noted almost two thousand years ago. After molting, he wrote, "the bird is abashed and moping, and seeks retired spots." But that's crediting Carl and Mr. Pickle with far too much self-awareness, as well as a nifty way of recasting as self-pity the vanity we humans habitually project onto a bird. They seemed only to realize, in a primal, cyclical way, that the ladies were done with them for the year, and there was no point in calling if no one was going to answer.

The pen for the season was split down the middle by three panels tacked together from wood scraps and chicken wire. Carl and Carlotta were on one side for a month until it became apparent that she had no interest in Carl and that, on the other side, Ethel had no interest in Mr. Pickle. It took a month to figure this out because the default position of a peahen is aggressive disinterest. Carl and Mr. Pickle exhausted themselves rattling feathers and prancing while Carlotta and Ethel lazily picked at bugs or things that might be bugs. Carlotta ignored Carl because his display was so pitiful. That was not unexpected. Ethel, I eventually decided, was too young: Like Carl, she'd probably come to us as a yearling. About a week after the racket started, I realized I'd made the wrong pairings and switched them up. Eudora would have been a much better name after all.

The first egg appeared in early May. It was a pale institutional beige and big, about two times the size of a chicken egg. Carlotta had left it in a divot near the waterer, like an errant golf ball she'd sliced into a sand trap. This was exactly the way nature intended peacocks to reproduce, but my excitement was embarrassing, even to me.

Louise was at work, but I called and left a message. "There's an egg! Call me." Then I called Valerie and left the same message. She called back three minutes later.

"She laid one? Oh, good. She's just getting started."

"Really? There will be more?" This came out like a yelp. "It's like the Sea-Monkeys have crowns!"

There was a beat or two of silence. "I'm sorry, sweetheart, what?"

I explained how I used to beg my dad to buy me Sea-Monkeys from the comic book ads. You just had to add water to get this instant family of Sea-Monkeys with crowns, and he always said no because they were just brine shrimp and I'd end up with a bowl of dirty water. But I kept asking and I finally got some and it was a bitter disappointment because they were just brine shrimp and in three days I had a bowl of dirty water.

In my analogy, this first peacock egg was like getting the real as-advertised Sea-Monkeys, which so rarely happens in life when one expects the astonishing. Inside that egg was a *real peacock* that would one day peck his way through the shell and grow into Mr. Pickle. I felt like I was eight again.

Valerie laughed at me, not really with me, but I could tell she appreciated my enthusiasm. We discussed the low probability of Carlotta sitting on a nest—she'd never sat before, and the pen didn't have the sort of underbrush she would instinctively nestle into—and also, I didn't want more peacocks. I decided to give all the eggs to Valerie to incubate, hatch, and then sell the chicks as she pleased. There was no point in them going to waste, and I wasn't doing any of the work. Besides, she'd been advising me for free; I figured the tip for the sand alone was worth at least a dozen Spalding black shoulder India blue eggs.

Carlotta laid another one the next day in a different spot, and two days after that, I found one in the sand beneath one of the roosts. "Perch bomb," Valerie called it when she picked up those three. "They'll just drop 'em."

Peacock eggs incubate for twenty-eight days at ninety-nine degrees, give or take half a degree, in a cabinet that keeps the temperature and humidity constant. They need to be turned at least once a day, either by hand or by machine, and on the tenth day or so, each one can be candled, which means shining a light through it to see if an embryo is developing. The infertile ones are discarded.

All of Carlotta's were fertile, and one of them hatched one evening in early June. Louise and I were on our way out the door to dinner when Valerie texted me a picture. He was gloriously ugly, like a teeny waterlogged duck, all legs and head, and I was sincerely giddy. I texted Valerie from the car and told her as much.

I'M SO GLAD, she texted back. HE NEEDED A LITTLE HELP GETTING OUT, BUT THAT'S OKAY. HE'S DOING FINE. I SPENT YEARS AND YEARS AROUND TRAGEDY AND DEATH, AND IT'S JUST SO NICE TO HELP BRING NEW LIFE INTO THE WORLD.

Chapter Twenty

There is an empty lot across from our house that used to be a meadow. The school down the street that owns it needed a temporary parking lot, so the trees were cut down and the brush was scraped away and most of the lot was covered with a layer of crushed and compacted stone. There is a narrow entrance through a stand of loblolly pines, and there is still meadow along the edges, tall grasses and maple saplings, bright splashes of buttercups and periwinkles in the spring and stands of goldenrod in the early autumn.

I'd taken to walking Tater there in the mornings, after Louise took Emmett to school on her way to her office but before the school bus came for Calvin. I'd stay there, usually until the bus came and went, before I let Tater lead me back across the street.

One morning in early October, while Tater was peeing on a sapling, my phone chirped with a notification from Google, the photo division. *Rediscover this day,* it read, as it does every so often. Like most people, I have a conflicted relationship with those notices. I appreciate the easy storage and occasional automatic flashback, but I'd also told Google everywhere I'd been and when, which is the sort of information no rational person should voluntarily surrender to anyone, let alone a corporation. Yet here we are.

The day I was rediscovering that morning involved a bright green field that I did not recognize and another, dustier one with shrubs and small trees planted in neat lines. I swiped through the photos, and a shining coil of fresh concertina wire cut across the foreground. Then there were people, men and women and children, a line of them two abreast and stretching over the field until it disappeared behind a distant tree line. This was familiar. The people were refugees, mostly from Afghanistan and Syria, which had collapsed into a barbarous civil war, and they were crossing from Croatia into Hungary, where they were herded onto a rundown train next to a rundown railway station. This had happened three years earlier in a tiny village called Zákány, and the train was taking them to another border town, Hegyeshalom, from where they would walk into Austria. Before it left, the train backed onto a side rail, where volunteers had twenty minutes to distribute water and sack lunches that contained three pieces of bread, two pieces of cheese, one banana, and a cookie or a candy bar.

Twelve hundred people got onto that train, and twelve hundred got on one earlier that day, and another later, and another after that. Hundreds of thousands of migrants were straggling into Europe in the summer of 2015, which had not been unexpected; the war in Syria had exacerbated ongoing crises of refugees from Afghanistan and Iraq. I had gone to Hungary to write about seventy-one of those people, four of whom were children. Smugglers they had paid for safe passage to Austria locked them in a refrigerated truck that used to haul processed chicken. The refrigeration unit no longer worked, but the truck was still airtight. Most likely, everyone in the back suffocated before they got north of Budapest. The smugglers abandoned the truck, and the migrants' bodies had started to decompose by the time an Austrian highway worker found them baking on the side of the A4 a day and a half later.

Millions of people fleeing war and famine can be seen either as a humanitarian crisis or, if you're a sociopath, as an opportunity for grotesque political theater. Viktor Orbán, the autocrat who commandeered the prime minister's office in 2010, chose the latter. He made a grand show of erecting a barbed-wire fence on the Serbian border to, in his telling, protect "Christian Europe" from an unchecked horde of Muslim invaders; and for a time, he corralled several thousand refugees inside Budapest's main train station, through which they had traveled unimpeded for months. Not coincidentally, trapping people in the station provided a stark visual of a large crowd of exhausted brown faces that a nationalist goon could use to justify his purported defense of a continent. None of those refugees wanted to remain in Hungary—it's a relatively poor country officially hostile to outsiders, and they were all waiting on outbound trains—but that was beside the point. Orbán claimed to be protecting the whole of Western Europe.

Except he wasn't. Aside from Europe not needing to be protected from exhausted and impoverished people whose immediate objective was simply not dying, Orbán wasn't stopping anyone from migrating westward. That's why I was in Zákány: The Hungarian authorities had merely diverted the refugees through Croatia and into a lonely, invisible border town. It was an illusion that made the lives of desperate people marginally more desperate while accomplishing nothing of substance.

I would not have remembered that day, or thought of refugees, if my phone hadn't reminded me. There is only so much unpleasantness one can comfortably hold in the easily accessible reaches of the brain. It's not willful amnesia. I remember the stories, most of them, in broad strokes because each in its time consumed an unreasonable proportion of my mental and emotional attention. They're macabre markers throwing shade on more significant memories. Calvin, for example, was born between sex traffickers in Moldova and prostitutes in Costa Rica, and he was a frog in the preschool play right after eleven men were killed on an exploding oil rig in the Gulf of Mexico, which I know because I flew home during the ensuing environmental catastrophe to watch him. Emmett was hard into his minerals-and-gems phase when he asked me to bring him some rocks from Arizona, where nineteen firefighters had burned to death, and he was still in it when I brought him an opal from Australia, from where the search for a missing airliner was being run. Louise is tethered to all of them, a through line. She is the one to whom I repeat the terrible stories, who patiently watches each project spiral into an obsession until, suddenly and abruptly, it no longer is. It must be exhausting.

I wondered if it was lonely, too.

Swiping through photos, I remembered a thread of humor, black as tar, that made me smile grimly and shake my head. Orbán once bragged that he had been able to transform Hungary into a one-party fascist-curious state by fobbing off critics with superficial gestures, feints toward democracy. "The dance of the peacock," he called it.

The squeak of brakes interrupted, the school bus sliding to a stop. The birds ignored the sound, but in the spring, at the beginning of breeding season, the noise set off Carl and Mr. Pickle something awful. That's how Calvin's friends discovered we had peacocks, from the noise; until the pecans leafed out they could see straight up the driveway, see flashes of blue breast in the distance. Some of them told Calvin we must be rich, having pea-cocks and all.

I don't think he ever tried to dissuade them, and I'm certain he never mentioned that the coop was built with garbage.

I requisitioned two more hens from Valerie, one white, the other silver pied, which, as opposed to regular pied, means at least sixty percent of the feathers are white. She didn't charge me. I'd given Valerie all of Carlotta's eggs that summer, thirty of them, all fertile, and she knew I needed the hens. My friends at the UPA recommended a four-to-one ratio of boys to girls, but a total of ten birds was lunacy. Four girls and two boys, a total of six, was acceptable.

Louise was in her office that evening, marking up a manuscript with scribbles only she could decode, turning the page sideways to make them all fit in the margins. I stood in the doorway, knowing I was interrupting, but it seemed important.

"Valerie's got those two girls ready for us."

She did not look up.

"Did you hear me? They're both white. They'll even have white toodles."

"Okay," she said, still scribbling, not looking up.

I sat down. "So I guess I'll start reframing the coop tomorrow."

She put her pen down. "You mean *expanding* the coop."

"Well, yes, obviously. They need more space."

Louise swiveled to face me. "Where will the boys' bikes go? The firewood? The wheelbarrow? All those bags of topsoil? And our bikes—"

"There's room. I'll push all that—"

"All *that*? You mean the stuff we actually use?"

"Yes. All that can fit in the front."

"Sounds like the birds are getting the whole barn," she said. It sounded more like an accusation than a statement.

"It's not the whole barn. And what do you want me to do? We need two more hens—"

"*Need?* We talk to the boys all the time about the difference between want and need."

She was making this sound like a choice, like I was being selfish. And I hated being compared to the boys, like I wanted Lucky Charms for dinner or something. "That's cheap," I told her. "I'm not a fucking child."

I knew that comment would sting. I stood up to walk away but stopped. I didn't want to prove her point. When I turned back around, she was studying me.

"Welp," she said. I noticed the tiniest curl of a smile. "It's cheaper than a sports car, I guess. And less obvious than dental implants . . ." She was choosing her words carefully. "Look, as long as *you* know this isn't just about peacocks." There was a gentleness to the way she said it that was unsettling, as though she saw something dangerously fragile in me that I could not.

But now I had six of them. And a bigger coop to build.

With the woodpile moved to one end, there was enough room to double the size of the pen. It was a straightforward expansion. The enclosure where Comet and Snowball had been locked up every night but not during the middle of the afternoon when foxes could stalk them in the yard would be removed, and I would frame wire walls across the front and on the new end. The roof sloped up steeply in that section, and the ceiling was almost twenty feet high at the back, which was excellent for the peacocks but required more wood and wire and time to seal up.

A truck dumped another eight tons of sand near the coop. Calvin and Emmett helped move half of it, but there was no rush—the new hens wouldn't be big enough to move in before January. I worked piecemeal on the enclosure for a couple of weeks, cutting a few boards here, tacking up a roll of chicken wire there, shoveling sand if I was feeling ambitious.

Ethel was endlessly curious, following me around with her eyes, cocking her head at different sounds, the double click of a staple gun, the grumbly whine of a circular saw with a fading battery. Her gaze eventually would catch mine, and I'd squat down by the wire and talk to her. I'd show her a tool and explain how it worked, or tell her what I was going to do with the rest of the afternoon, and then Carlotta would wander over and Carl and Mr. Pickle would follow because clearly something interesting was happening that might involve blueberries.

It was possible that I enjoyed the peacocks more, collectively, in the weeks after the summer heat faded. There was no notable change in the girls from season to season; neither of them got broody in the summer, so they remained the same docile, pleasant company year-round. The boys, on the other hand, were at their most absurd. Their trains had fully molted, which reduced their overall length to less than half of their springtime prime. Instead of a bejeweled carpet flowing behind them, each had his tail feathers, gray as a midcentury chalkboard, cantilevering off his rump. Without the train for balance, the colors that remained—the cobalt breast and neck, the green-gold saddle between the wings, Mr. Pickle's blue-black shoulders—were a preposterous mess. There was less to distract the eye from the legs and feet, which could at least be respectfully mistaken for the deadly appendages of a savage predator had they not been attached to a stubby, gaudy creature that appears less like an actual bird than a vandalized scrap of yard art. It was as if each had been demoted from peacock to a lesser, goofier species. A peacock without his train feathers has been stripped of the one thing that confirms he is, in fact, a peacock.

And yet a peacock in autumn does not care. He does not even admit that it has happened.

When his incoming coverts are barely more than nubs, a peacock will flex and rattle his tail feathers the same as he did in May or July, those glorious golden months when he was in full flower and his eyespots hovered in a swaying field of copper and turquoise and people gaped and convinced themselves, some of us, that this was a magnificent bird, majestic and mythic and quite possibly immortal. In the autumn, the presentation was ridiculous, gray feathers buzzing behind miniature eyespots and fishtails. But a peacock does not know that. He does not consult a mirror. He raises his embryonic train out of instinct and habit, but I prefer to consider it a defiant gesture, a bird quietly raging against the cruel caprice of nature.

I took a picture of Carl displaying, as it were, for my friend Ted. We met decades ago at an alternative weekly that no longer exists; Ted was the receptionist before he got promoted to selling advertising space to escorts and porn shops, which was a way that newspapers made money before the Internet. He quit that, went to business school, got a good job at a bank and then better jobs at better banks until the last one laid him off. He was suffering an uncomfortably long period of middle-aged unemployment. A PEACOCK IN AUTUMN, I texted.

He texted right back. YES, WE ARE.

THE FEATHERS GROW BACK!

UNTIL THEY DON'T, he replied.

Why Peacocks?

· · ·

One of the new hens was limping. That's what we called them, The New Hens. Calvin and I had brought them home in January, but after the wrongheaded naming of Carlotta, none of us could commit, and we got frustrated trying to think of two perfect names, so we just called them the new hens as a placeholder.

The pure white one might have had some Spalding in her, but I was only guessing by her crest, which seemed taller and tighter than even Carlotta's. The silver pied, mostly white but smudged with charcoal, seemed to be all India blue. She was the one limping. On one of her feet was a dark lump that could have been bumble-foot, a staph infection that ground-dwelling birds are known to get. I couldn't get close enough to confirm either way—the new hens were skittish—but I texted pictures and a short video to Burkett.

He called me a little while later. "That's a pretty serious limp," he said. "But I can't tell what's on her foot. Can you get a better picture? You're probably gonna have to catch her."

I did not want to catch her. I did not handle any of the peacocks. I liked sitting with them, and I was pleased that they were comfortable enough to line up for treats. But they are not cuddly animals. I had no desire to stroke their feathers and even less to agitate one. But what choice was there? Unchecked bumblefoot can kill a bird.

Uncle John's yellow net was folded on a shelf. I positioned Calvin in one corner in case I needed an extra set of hands, stretched the net out in front of me, and turned toward the pied girl. I hadn't caught a peacock since Carl, and he'd been hobbled by poison. That was, what, eighteen months ago? Had these birds been here that long already?

I herded the pied one into a corner, moving deliberately, convinced that I wouldn't spook her if I took my time. It did not occur to me that a bird would consider a looming net a threat at any speed. Dragging it out just gave her time to plan. I'd forgotten how Valerie had grabbed Carl, the lightning snatch that didn't give him time to resist.

The bird was against the wall, two feet from me. And then she wasn't. There was only an explosion of white and gray, head high and coming at me fast as an airbag. I ducked, twisted my face away. One of her feet caught me behind the ear, and three toes raked across my scalp.

The pied girl was on the other side of the pen, standing calmly by herself. The other birds had all moved to the periphery, as if watching a street fight.

"Well, that was stupid," I muttered. Calvin was staring at me, slack-jawed. "Am I bleeding?"

"No." He blinked. "Maybe."

I could feel blood dripping down my neck. I touched my ear, examined it. There was a warm wetness in my hair, but nothing was missing. A couple of deep scratches. I dragged my fingers across my shirt, wiped off the blood.

"Uh, yeah," Calvin said. "I think you're bleeding." He seemed to be deciding whether this was frightening or funny.

"She's just scared, buddy," I said. "She wasn't trying to hurt me, only trying to get away. I wasn't fast enough."

"Oh, I know," he said, rolling his eyes. *Ah, good. Funny.*

I unfurled the net and went directly at her with a quick, unbroken stride. She dodged right, then left, but I had her corralled. She flew up and I tossed the net over her. She dropped like a stone. I got to my knees, pressed her against me with one hand, and seized her legs with the other. I flicked at the black spot on her foot with my thumb. It fell away, leaving a bare, clean, uninfected foot. Poop.

Calvin helped me untangle the net, and the pied hen hopped away. She was still limping, but it wasn't because of an infection. I let out a long breath. *Just poop.* I hadn't realized until then how much I'd feared an infection: I wasn't sure I could afford another extended stay at the Birdie Boutique. Two hens were supposed to make things better, not cause more trouble.

That's the problem with trying to create your own Eden. There's always a serpent hiding nearby, waiting to slip in and wreck the place.

All the questions I'd never asked Danielle were gnawing at me. Not in a bad way, necessarily. I liked my peacocks. They, or one of them, had been more expensive than I'd expected, but they were still pretty to look at, still attentive listeners, still a distraction, a personal counterweight to the professional. They were comfortable with me, and I took some strange pride in that fact.

I met Danielle for coffee in a shop not too far from her farm. She told me about her grandfather and how the family came to live on those acres with horses and peacocks. She was an ocularist, too, hand-painting artificial eyes. There are only three in North Carolina, she told me, and I remarked that I wouldn't expect there to be even that much of a demand. "More than you know," she said, between diseases and injuries. "You know that Christmas movie where the kid wants a BB gun and everyone tells him he'll shoot his eye out? Yeah, that happens." A battered woman tends to lose her left eye because most men are right-handed. Also, blue eyes are the most difficult to paint.

"I don't think I've ever seen someone with an artificial eye," I said.

"If I do my job right," she said, "you won't."

We talked for a while longer before I brought up Burkett. She knew him. Not well, but he lived a few miles up the road from the farm, had bought some hay from her once or twice. I told her about Carl, how he'd had his blood chelated like Keith Richards. She smiled, somewhere between sympathetic and amused.

"So here's the thing," I said. "It's kind of funny, really, and it doesn't really matter, but I have to ask. Remember how you told us about the owl?"

She nodded.

"Okay, so I told Burkett about that, told him the whole story."

She nodded again.

"And, um, well, he said that didn't happen. *Probably* didn't happen. That an owl wouldn't bite off the head and leave the rest."

"He said that?"

"Yeah. Like, right after I got them, too. I mean, I looked it up and it does happen sometimes, owls do bite off heads. But I've learned a lot of other things about peacocks, too, since then. So there was this one thing Burkett said that I was wondering about. He said—wait, let me make sure I've got the words right—he said, 'Danielle's been trying to get rid of those goddamned birds for years.'"

She looked at me, held a blank expression for a few seconds.

Then she laughed. Out loud, in a coffee shop. Laughed and laughed.

"He really said that?" She laughed some more. "Okay, I was finding some headless birds. But one of those little sons of bitches would sit in front of the truck, right in front of the silver bumper, and pick a fight with himself. There'd be all these bloody spots on the bumper where he'd been pecking. And then he'd get up on the *hood*, all triumphant and dominant, like he won."

"Yeah, I've heard those stories," I said.

"They just became a liability. I mean, I can't afford to keep re-painting my truck."

I laughed with her. All the rest made sense, the price drop, the speed with which she was bagging up birds for sale, and, finally, the reason she would want to get rid of them. I told her the birds were locked up now, that they had a spacious coop and seemed happy. And that I didn't really care whether there was an owl. I liked having peacocks, and if we'd known about the bloody bumper and the scratched hood from the beginning, we probably still wouldn't have walked away.

I ran into Burkett a few nights later at the Sportsplex, where there's a gym and a couple of pools and an ice rink. I helped coach Calvin's hockey team, though only in the most generous sense of the term. Mostly, I collected pucks and moved nets and did whatever the real coach asked me to do. I was never more than a mediocre player, but I worked at ice rinks during high school and college, driving the Zamboni and sharpening skates and teaching kids to penguin-step across the ice. The rink was comfortable, familiar without being nostalgic, and if I was going to be there two nights a week, I'd rather be on the ice than sitting on a bench. They're metal. Very cold.

Burkett was sitting outside one of the group exercise rooms. I didn't recognize him right away, just saw a man, thin and bald and dressed in loose black clothing, glancing at me and looking away and then glancing again, the way people do when they're certain they know you but can't for the life of them figure out why. I was doing the same thing.

"Dr. Burkett," I said, plopping myself next to him. We were both out of context, which was why it took us a few seconds. "What is my favorite avian veterinarian doing in this place?"

"Yoga," he said. I would not have previously pegged him as a yoga man, but it made sense in this environment. He had the build for it. Also, he was holding a yoga mat.

I asked about the bird business and he asked about the peacocks. I told him the silver pied hen had gradually lost her limp. My new theory was that she'd come down hard from a high roost and pulled a muscle or some such. I told him about a coop-building seminar at the UPA convention, where I'd learned there should be three feet of glide space for every foot of descent so birds don't hurt themselves making a steep drop. The planks I'd mounted ten feet above the sand did not have anywhere near the preferred thirty-foot landing path.

Burkett said that sounded reasonable.

"I'm fucked, aren't I?" I smiled when I said it.

He chortled. I remembered the sound from the day I met him. "What do you mean?"

"The peacocks. I'm stuck with them."

He raised a curious eyebrow.

"I mean, I can't let them out. Not because they'd fly away, but because they'll menace the neighbors. And if they did fly away, they'd be a menace somewhere else. So I've got them locked up in this fortified cage, and now you can hardly see them from the outside. And they don't even care. They don't know that we want to look at them, and they don't want out. They're perfectly content lounging around in their giant garbage condo."

"Why would they want out?" he asked.

I scrunched my brow. The question seemed nonsensical. I had assumed almost any caged animal would prefer not being caged, especially one who had roamed free until someone shoved him into a feed sack and zip-tied his legs. Mr. Pickle and Carl and Ethel weren't industrial farm chickens—they were birds of Greek gods and biblical traders, of myth and legend, fantasy and magic. They should want out, I supposed, because that's what I would want, would *demand*, in their position. That they did not was befuddling.

"I'm sorry," I said. "What?"

"Why would they want out?" he repeated. "They've got plenty of space, they've got food, they've got water, and they're safe. That's a pretty good setup."

"Yeah," I said, "but they're birds."

Burkett studied me for a moment, as if he was waiting for me to realize my own epiphany. He nodded slowly. "That's right," he said. "They're birds. They're just birds."

Chapter Twenty-One

Okra started dying one night in November. She'd always been a sickly cat, but in ways that were gross and idiosyncratic, not debilitating. She was allergic to mosquitoes, so half the year we were rubbing ointment on her scabby nose and ears. She'd pull out the fur on her back sometimes, too, give herself a Mohawk that ran to the base of her tail, and she went through periods of pissing and spraying on random pieces of furniture. She wasn't affectionate until she was, and then she was clingy and demanding, almost nasty about it. There had always been at least one cat, and usually three, in my house since I was seven years old, and Okra was the only one who made me believe, as a Scottish study suggested, that cats would try to kill us if they were bigger.

This was a different kind of sickness, acute and rapid-onset. We had no idea how old she was because she'd arrived fully grown eleven years before, but suddenly she was decrepit. She wobbled out of a bedroom and bumped into a wall, though not hard enough to hurt herself because she was moving so slowly. "Okra?" I said. She bumped the wall again. I bent down and rubbed behind her ear to get her attention, and she turned toward me. She looked like she'd crawled off a black-velvet painting: Her pupils were big as nickels, fully dilated, the way they would be if she were hunting in the barn on a moonless night. And she was frighteningly thin. She'd been losing a little weight over the months, but I hadn't realized until I picked her up that she was bony, almost skeletal.

I set her on the chair near the front door where she liked to sleep. I waved two fingers in front of her face. Nothing. She was blind. I snapped my fingers next to one ear, then the other. Deaf, too.

I found Louise in the kitchen, then peeked into the dining room to make sure the boys weren't close enough to overhear. "I think Okra's dying," I said.

"Again?" she said, without shifting her attention from the onions she was dicing. "There's always something wrong with that cat."

"No, really. She's *dying* dying."

She stopped dicing and looked at me, trying to read my face. "You're serious," she said, putting the knife down and wiping her hands on a dishtowel. "Where is she?"

I nodded toward the front hall and followed behind her. Okra was still on the blue chair, shrunken and still. "Oh no," Louise whispered. She bent down and stroked the base of Okra's neck, her fingertips slipping into a divot between the cat's shoulders. She glanced at me, clearly unnerved by Okra's rapid decline. "Do the boys know?"

"Not yet. I just found her."

"Let's not rush it," she said, turning back to the cat, softly touching a spot behind her ear. "You poor sick girl."

There was a familiar tenderness in her voice. Okra was a difficult cat on her best days, mewling and destructive on her worst, and a constant allergen to Louise regardless. We hadn't adopted her so much as she just never went away, and then somehow, without anyone ever planning it, we were taking her to the vet and paying for creams and ointments and special kibble so she wouldn't get fatter. The boys adored her, but they weren't the ones scrubbing pee stains out of the rugs. On balance, Okra could be fairly described as tolerable.

Until she was dying. Nothing stirs forgiveness and softens perspective like imminent death. It's instinctual: Calvin and Emmett were never more fond of Carl than when he was bubbling with poison. Whether that instinct is born of empathy for another's misery, a fear of one's own impending loss, or some combination of the two is immaterial. What matters is that it exists. Louise and her father were closest in the last, irradiated years of his life, when his cancer couldn't be fixed but they each could somehow make it better. She is finely tuned to the suffering of others, and her gentleness is directly proportional to their vulnerability.

Including, I knew, to that of a cat who made her wheeze. She slipped away to the kitchen and returned with a shallow ramekin of water and a tiny scoop of finely chopped tuna. Okra nibbled at it from a spoon Louise held up to her mouth, then lost interest and returned to staring wide-eyed at nothing. We, in turn, stared at the cat, quietly calculating her odds of midterm survival.

I didn't hear Emmett coming until he ducked past me and sat on the floor so he was eye-level with Okra.

"What's wrong with her?" he asked, sounding more curious than worried.

"I don't know, pup. I'll get her to the vet in the morning." I doubted she was going to make it until morning, but he didn't need to know that. Okra was already here when he was born. He'd had a snake for two months and chickens for ten months, but he'd had a cat his whole life.

"Can she see?"

"No, I think she went blind," I said. "And probably deaf."

"Why?" He was looking at me, expectant. "Is she gonna be okay?"

I took a breath, bought a second to think. *Yes* would be a lie, *no* would be crushing and, possibly, incorrect; the boy had already witnessed a peacock brought back from the edge of death. I punted. "Maybe. I'll get her to the vet in the morning."

Emmett nodded, then turned back toward his cat. He stayed on the floor for almost an hour, stroking her paw and feeding her bits of tuna.

Okra would probably spare him in her killing spree.

She was still alive the next morning, though she was too weak to resist when I slid her into the cat carrier.

The vet disappeared into the back with my emaciated cat. I waited in a small exam room, reading pamphlets about heartworms and fleas and vaccines for feline immunodeficiency virus. We'd had another cat years before, Pasha, who'd come to us already infected with FIV. We had to put him down a few months before all of Otis's organs slid out of place.

There was a knock on the door, followed by the vet stepping inside. She was about Louise's age, and she had the same soft touch for bad circumstances. "Okra's still in the back," she said, "getting some fluids. How are you?"

I told her I was fine, and she told me Okra's blood pressure was dangerously high and that an extreme spike had detached her retinas. The doctor didn't think the cat was deaf but, rather, that she might have been ignoring me because she was disoriented. There were medicines to get her blood pressure stabilized and subcutaneous liquids to get her rehydrated and some other pills to do another thing.

"So this is something we should be treating?" I asked, trying to be delicate with that last word.

The vet paused but didn't break eye contact. "It's all treatable," she said with what sounded to me like a practiced optimism.

I'd put the emphasis on the wrong word, on *treating* rather than *should*. Really, I'd been asking if it would be kinder to put her down, but instead I'd presented myself as a person who would take extraordinary measures to save a cat. Which we'd already done for Otis. And a bird no one could even pet.

Okra came home from the vet with a bag of pills and syringes preloaded with an appetite stimulant to squirt down her throat twice a day. The syringes were easy, but she'd always been impossible with pills. She didn't have much fight in her, but she could still spit a pill like a kitten. Louise ground them into a powder that she sprinkled on top of a dollop of the wettest cat food we could find. Okra didn't eat past the top layer.

Her pupils shrank and she seemed to regain at least some of her sight, shadows and shapes, light and dark. She did not otherwise get better. She stayed on the blue chair most of the time, deteriorating but not uncomfortably so for a few days. After a week, though, she was clearly ready to be done with it all. I made another appointment with the vet.

That evening, Emmett was on the floor in front of her again, like he had been every night. He had two fingers on her paw, and she was purring, a raspy, rattly sound, as if cartilage was coming loose inside.

I sat down beside him, put a hand on his back. "She likes you being here," I said. "You make her feel safe."

We stayed there, the three of us, silent except for Okra's grinding purr until he asked when we could take her back to the vet.

"Tomorrow morning, pup. I made an appointment."

"What do you think they'll do for her?"

I was glad he wasn't looking at my face. I wasn't sure if I should say it, and I took a few seconds to decide. "I don't think she's coming home, pup."

He kept his eyes on his cat. "What do you mean?"

He knew what I meant.

"She's dying, pup."

"But they can make her better."

"Maybe, but I don't think so. They can keep her alive, but that's not the same thing as better." I paused to let him digest that. "Do you think she's happy right now?"

He stroked her cheek with one finger. Okra stared back without appearing to see. I told him the story again of how she showed up in the okra patch one day and lived in the barn and wouldn't go away so we took her in. We talked about the mice she left on the back steps for us when she was young and feisty, and we decided that, all things considered, she'd had a long and happy cat life. She had food and water and safety and people who loved her.

Emmett started softly crying for his cat. I stayed with him, not saying anything because there wasn't anything left unsaid. I wasn't sure that I'd gotten any better at discussing death with him, but at least we could trudge through it together. Maybe that was all that mattered.

After the boys were in school, Louise bundled Okra in a towel and held her on the drive to the vet. A crate seemed needlessly cruel at the end. We held her while she drifted off, and took her home wrapped in plastic and sealed in a box that we could bury next to Cosmo.

I picked up Emmett from school early that afternoon. He got into the car with a forced hopefulness. "Is Okra home?"

"No, pup. I'm so sorry."

He slumped into the seat next to me, sighed. He did not seem surprised. "That's okay. I didn't think she would be."

It was a short drive home, no longer than five minutes, and he was quiet until we were at the back door. "Do you think it hurt?" he asked.

No, I told him, I was sure it didn't. "The doctor gave her a shot, and she fell asleep with me and Mom."

"Do you think she knew? That she was dying?"

"I don't think so. I think she knew she was tired and not hungry and couldn't really see and some parts hurt. But I don't think animals have the same concept of death that we do, that one day we won't be here."

He considered that for a moment, then nodded. "That's good. She wouldn't be scared if she didn't know."

Tater heard us on the porch, started with his insistent, harmless bark. "Wait a minute, dog," I said through the door, and turned back to Emmett. His eyes were damp again. "Did you want to talk some more?"

He shook his head. "Not now," he said. "I just wish all my pets didn't die."

He opened the door before I could answer. Tater shot out, wiggling and panting as if we'd been gone for days.

"We've got Tater," I said, scooping him up and depositing him in Emmett's arms. "He's young and healthy. And we've got peacocks—"

"Those are yours, Dad," Emmett said.

"Yes, I know. I see that more clearly every day," I said slowly. "But they'll end up yours. I hear they're supposed to live forever."

Acknowledgments

When you write about other people for long enough, it becomes possible, even habitual, to find a narrative in even the most ordinary-seeming lives. Except, perhaps, your own. For that, you sometimes need the sharp eye of an excellent editor, who in this case would be Sean Manning. Over Chinese food and cocktails one night, he began teasing out of me the story of my strange birds; he saw something I could not, and for that I will always be grateful. He shaped the manuscript and took out the lousy parts, made sentences and paragraphs and chapters so much better, and he did so with a tremendous amount of patience.

David Black has been a fierce advocate and loyal friend for more than twenty years. It's a privilege to be represented by him and to work with everyone at his eponymous agency, especially Joy Tutela, Sarah Smith, and Ayla Zuraw-Friedland.

Any errors, and accompanying embarrassment, are mine alone. But there aren't any because Julia Ellis, a tenacious researcher, double-checked every fact in the preceding pages. Thanks also to Jonathan Karp, Dana Canedy, Richard Rhorer, Elizabeth Breeden, Tzipora Baitch, Brianna Scharfenberg, Alison Forner, E. Beth Thomas, Yvette Grant, and the rest of the Simon & Schuster team.

Tanja Vujic sent the text message that brought Ethel, Carl, and

Acknowledgments

Mr. Pickle to the garbage coop. Jonathan George, Jason George, and the staff at Barnes Supply Company helped me keep the birds and Tater well-nourished; and Christine Bristor recommended we buy the peeping puffballs who grew up to be Comet and Snowball. Cameron Alworth and Berit Brown keep this place looking like a little farm, and they tolerate the summertime noise. Andrew Ovenden first introduced me to a domesticated bird, and Paula Scatoloni, I realize now, began pulling at the first narrative threads years ago.

Roslyn Dakin and Jessica Yorzinski indulged my early, ill-informed questions and kindly explained their research to me. Lisa Schubert was a delightful and welcoming guide to St. John the Divine, and Kathy Kerran generously shared a trove of stories about the peacocks at the Los Angeles Arboretum. Rebecca Halpern introduced me to Palos Verdes, and Mary Jo Hazard, Kirk Retz, and Jack Alexander filled in a lot of the background, as did Ana Bustillos of the SPCALA.

Ian Moir of the Fire Station Creative, Lisa Edwards of Dunfermline Delivers, Jack Pryde of Discover Dunfermline Tours, and Frank Connolly were excellent hosts. Judge Andrew T. Park dug through old files to help a stranger with a weird request. Justin Diggs delivered a whole lot of sand, Mary Thacher and Holly Rogers let me in on a joke, and Sadie Fraleigh, DVM, was kind and compassionate to an old cat and two sad people. I'm indebted to all of them.

I'm also lucky to live in a place with so many other writers who are unfailingly supportive and generous, including Bronwen Dickey, David A. Graham, Haven Kimmel, Barry Yeoman, and Jason Zengerle. I'm grateful, too, to Eric and Lisa Guckian, John and Susan Haws, and Ted Miller and Bernadette Carr. Thank you.

Calvin and Emmett are never-ending sources of wonder and joy—unfailingly kind, preternaturally curious and, at times, supremely patient. I am more proud of them every day.

Acknowledgments

Finally, one reason to fall in love with another writer, to borrow a sentence, is self-evident: These words would not have made it to paper without the patience, counsel, and tireless creative labors of Louise Jarvis Flynn. The story wouldn't exist without her, obviously, but it might not have been written, either. Louise is my first, best editor, in this and everything, and will be always. Her talents are boundless, as is my love for her.

Notes on Sources

This is a work of nonfiction and memoir. All the people are real, as are the memories, and wherever possible those memories have been checked against those of the people involved. Personal interviews accounted for much of the remaining research, and those conversations are apparent throughout the text; some sources, such as books and newspapers, are noted, as well.

But space and clarity preclude noting every source directly in the text, whether a helpful human or an academic paper, internet database or newspaper archive. In chapter one, for instance, I incorporated the transcript of an oral history interview with Lee "Shorty" Barnes conducted by Michael Smith on June 13, 2000; and chapter two uses information from the *Journal of Ophthalmic Prosthetics*, fall 2015 issue. The Oscar Wilde quote in chapter four is from "The Picture of Dorian Gray" (*Lippincott's Monthly Magazine*, July 1890).

Details in chapter five were gleaned from IMDB.com, nostalgia.com, and the archives of Reuters, the *Los Angeles Times*, the *New York Times*, and UPI. Chapter six was informed by the web archives of the Cathedral Church of St. John the Divine and the work of composers Paul Winter, Jim Scott, and Paul Halley. The quote on page 75 is from P. Thankappan Nair's "The Peacock Cult in Asia," published in 1974 by Nanzan University in *Asian Folklore Studies*, Vol. 33, No. 2.

For chapter eight, Roslyn Dakin analyzed observational data of peacocks displaying; and Nathan Hart, who is the head of the Department of Biological Sciences at Macquarie University in Sydney, Australia, explained the physiology of avian eyesight.

The letters excerpted in chapter nine were retrieved from the Darwin Correspondence Program at the University of Cambridge. John Ruskin's quote is from *The Stones of Venice*, Volume One, published in 1894. I also used information from Greenpeace East Asia, the archives of *Beijing Review* and Bloomberg, and the April 1966 issue of *The Auk* (volume 83, issue 2). An enormous number of academic studies informed the general background and some specific details, most notably the work of Roslyn Dakin of Carleton University in Ottawa, Suzanne Amador Kane of Haverford College in Pennsylvania, Robert Montgomerie of Queen's University in Ontario, Jessica Yorzinski of Texas A&M University, Michael L. Platt of Duke University in Durham, and Jian Zi of Fudan University in Shanghai.

Chapter eleven drew upon the work of Kristopher Poole, a zooarcheologist at the University of Sheffield, particularly his chapter in "Extinctions and Invasions: A Social History of British Fauna," edited by Terry O'Conner and Naomi Sykes (Windgather Press, 2010). Alan Bergo of Forager/Chef provided the culinary basics; and Suzanne Turner Associates compiled a timeline of Fredric Church's Olana estate.

I drew upon research by photographer, journalist, and author Vicki A. Mack in chapter thirteen, as well as records kept by Mary Gliksman and the archives of *Los Angeles* magazine, particularly the January 2016 story "Who's Been Killing the Feral Peacocks of

Palos Verdes?" by Mike Kessler. The California Digital Newspaper Collection, a project of the Center for Bibliographical Studies and Research at the University of California, Riverside, was invaluable.

In chapter fourteen, the damage peacocks can do to a crop field was quantified in a 2018 study by Suresh K. Govind of the Forest Research Institute and E. A. Jayson of Christ College that was published in the journal *Indian Birds*. Chapter fifteen would not have happened without the assistance of *Dunfermline Press* reporter Gemma Ryder, peafowl warden Suzi Ross, and local historians Jack Pryde and Frank Connolly.

My neurosis about the world's dwindling supply of sand in chapter sixteen was stoked by, among others, Vince Beiser in *Wired* ("The Deadly Global War for Sand," March 2015); David Owen in the *New Yorker* ("The World is Running Out of Sand," May 22, 2017); and Harald Franzen writing for Deutsche Welle ("Could we run out of sand?" November 28, 2017). Brad Legg and his son Brandon grounded me in peacock genetics and breeding for chapter seventeen.

The California Digital Newspaper Collection was essential in chapter eighteen, as were the archives of *Modern Game Breeding and Hunting* and *Aviculture*. And the notion in chapter twenty-one that cats would kill us if they could is a (perhaps glib) extrapolation from a 2014 University of Edinburgh study published in the *Journal of Comparative Psychology*.

Selected Bibliography

al-Kisa'i. *Tales of the Prophets*. Translated from the eleventh-century original by Wheeler M. Thackston Jr. Chicago: Kazi Publications, 1997.

Anand, Anita, and William Dalrymple. *Koh-i-Noor: The History of the World's Most Infamous Diamond*. London: Bloomsbury Publishing, 2016.

Arakelove, Victoria, and Garnik S. Asatrian. *The Religion of the Peacock Angel: The Yezidis and Their Spirit World*. London: Routledge, 2014.

Attar. *The Conference of the Birds*. Translated by Shoaled Wolpé. New York: W. W. Norton & Company, 2017.

Bergman, Josef. *The Peafowl of the World*. Hindhead, England: Saiga Publishing, 1980.

Bland, Bartholomew F., ed. *Strut: The Peacock and Beauty in Art*. New York: Fordham University Press, 2014.

Brunner, Bernd. *Birdmania: A Remarkable Passion for Birds*. Vancouver: Greystone Books, 2017.

Carnegie, Andrew. *The Autobiography of Andrew Carnegie*. Boston: Houghton Mifflin, 1920.

———. *The Gospel of Wealth*. Boston: *North American Review*, June 1889.

Chopra, Praveen. *Vishnu's Mount: Birds in Indian Mythology and Folklore*. Chennai, India: Notion Press, 2017.

Glasscock, Carl B. *Lucky Baldwin: The Story of an Unconventional Success*. Reno: Silver Syndicate Press, 1993.

Hansen, Waldemar. *The Peacock Throne: The Drama of Mogul India*. New York: Holt, Rinehart and Winston, 1972.

Hazard, Mary Jo. *The Peacocks of Palos Verdes*. Los Angeles: Donegal Publishing Company, 2010.

Ingraham, Corinne. *The Peacock and the Wishing-Fairy and Other Stories*. New York: Brentano's, 1921.

Jackson, Christine E. *Peacock*. London: Reaktion Books, 2006.

Latham, John. *A General History of Birds, Vol. III*. London: Henry G. Bohn, 1823.

Lawler, Andrew. *Why Did the Chicken Cross the World? The Epic Saga of the Bird that Powers Civilization*. New York: Atria Books, 2014.

Mack, Vicki A. *Frank A. Vanderlip: The Banker Who Changed America*. Palos Verdes Estates: Pinale Press, 2013.

McAdam, Pat, and Snider, Sandy. *Arcadia: Where Ranch and City Meet*. Arcadia: Friends of the Arcadia Library, 1981.

Merrill, Linda. *The Peacock Room: A Cultural Biography*. New Haven: Yale University Press, 1998.

Montgomery, Sy. *Birdology: Adventures with a Pack of Hens, a Peck of Pigeons, Cantankerous Crows, Fierce Falcons, Hip Hop Parrots, Baby Hummingbirds, and One Murderously Big Living Dinosaur*. New York: Free Press, 2010.

Morales, Becky, Ernie Morales, and Evie Ybarra. *Images of America: Rancho Sespe*. Charleston: Arcadia Publishing, 2017.

Moyle, David. *Living with Peacocks*. iUniverse, 2006.

Nicoll, Fergus. *Shah Jahan*. New York: Penguin Viking, 2009.

Selected Bibliography

Pauly, Thomas H. *Zane Grey: His Life, His Adventures, His Women.* Champaign: University of Illinois Press, 2015.

Prum, Richard O. *The Evolution of Beauty.* New York: Doubleday, 2017.

Roberts, Michael. *Peacocks Past and Present.* Devon, England: Gold Cockerel Books, 2003.

Ryan, Michael J. *A Taste for the Beautiful: The Evolution of Attraction.* Princeton: Princeton University Press, 2018.

St. John, Percy B. *The Young Naturalist's Book of Birds: Anecdotes of the Feathered Creation.* Sherbourn Lane, London: Joseph Rickerby, 1838.

Snider, Sandy. *Historic Santa Anita.* Arcadia, CA: California Arboretum Foundation, 1976.

Sopa, Geshe Lhundub, Leonard Zwilling, and Michael J. Sweet. *Peacock in the Poison Grove: Two Buddhist Texts on Training the Mind.* Boston: Wisdom Publications, 1996.

Taillevent. *Le Viandier.* Translated by Jim Chevallier as *How to Cook a Peacock.* North Hollywood: Chez Jim, 2004.

Webster, Noah. *History of Animals; Being the Fourth Volume of Elements of Useful Knowledge.* New Haven: Howe & Deforest, 1812.